Uncertainty Modelling and Quality Control for Spatial Data

Uncertainty Modelling and Quality Control for Spatial Data

Edited by
Wenzhong Shi
Bo Wu
Alfred Stein

CRC Press
Taylor & Francis Group
Boca Raton London New York

CRC Press is an imprint of the
Taylor & Francis Group, an **informa** business

CRC Press
Taylor & Francis Group
6000 Broken Sound Parkway NW, Suite 300
Boca Raton, FL 33487-2742

First issued in paperback 2019

ISBN-13: 978-1-4987-3328-1 (hbk)
ISBN-13: 978-0-367-37714-4 (pbk)

Visit the Taylor & Francis Web site at
http://www.taylorandfrancis.com

and the CRC Press Web site at
http://www.crcpress.com

Contents

Section I Uncertainty Modelling and Quality Control

Section II Uncertainties in Multidimensional and Multiscale Data Integration

Preface

Data quality and uncertainty issues are essential for the development of geographic information science (GIS). Uncertainties always exist in GIS and are likely to propagate to outputs of spatial analysis and products. As a supporting science for spatial decision making, GIS best embeds uncertainty information for users to determine fitness of use for their applications. Therefore, uncertainty-related research is a very important research area in geographic information science.

The new developments in GIS in recent years, such as new insights into uncertainty modelling and quality control, multidimensional and multiscale data integration, national or regional spatial data products, and new spatial data mining methods, bring new challenges in spatial data quality research. This book provides the latest developments of uncertainty modelling and quality control of spatial data in these important aspects. The book is composed of 4 sections including 16 chapters of original research and developments in uncertainty modelling and quality control of spatial data. The book is organized as follows. It starts with uncertainty modelling and quality control, which is followed by uncertainties in multidimensional and multiscale data integration. Then comes quality control for spatial products. The book closes with uncertainties in spatial data mining.

The first section has four chapters (Chapters 1 to 4), covering review and new developments of uncertainty modelling and quality control of spatial data. The second section has four chapters (Chapters 5 to 8), covering data quality in multidimensional and multiscale data integration. Multidimensional and multiscale data integration has been emphasized more in recent years, and the data quality issues related to it have seldom been investigated in the past. The four chapters in this section provide new discoveries and knowledge in this aspect. Along with the production of national or regional spatial data products in recent years, uncertainty issues related to these products have become serious. The third section includes five chapters (Chapters 9 to 13) and addresses uncertainty issues related to spatial products from various aspects. As tremendous spatial data products are produced or underproducing in developing and developed countries, the chapters in this section will be helpful to understand the uncertainty issues in national or regional spatial products and to further improve the data quality of spatial products. The last section has three chapters (Chapters 14 to 16) and addresses the latest developments of uncertainty issues in spatial data mining.

As the next step in the current theoretical development for describing uncertainties in spatial data and analysis, this book provides new methods for analyzing uncertainties in various GIS applications. It brings the latest insights into spatial uncertainty modelling to readers in academic, industry, and government domains.

This book is supported by the Ministry of Science and Technology, China (project 2012BAJ15 B04), the Hong Kong Polytechnic University (project 1-ZE24) and National Administration of Surveying, Mapping and Geoinformation, China (Scheme for Technological Leading Talents). We would like to thank Rong Kou for her great efforts in assisting in the preparation of this book. Also, we would like to gratefully thank Taylor & Francis/CRC Press for their efforts and cordial cooperation in publishing this book.

Wenzhong Shi
Department of Land Surveying and Geo-Informatics
Hong Kong Polytechnic University, Hung Hom, Hong Kong

Bo Wu
Department of Land Surveying and Geo-Informatics
Hong Kong Polytechnic University, Hung Hom, Hong Kong

Alfred Stein
University of Twente, Enschede, the Netherlands

Editors

Wenzhong Shi is head of department and chair professor in geographic information science (GIS) and remote sensing for the Department of Land Surveying and Geo-Informatics, Hong Kong Polytechnic University, Hung Hom. He earned his BEng and MEng from Wuhan Technic University of Surveying and Mapping (now Wuhan University), Wuhan, Hubei Province, in 1985 and 1988, respectively, and his doctoral degree from University of Osnabrück in Vechta, Germany, in 1994. His current research interests are in the areas of GIS and remote sensing, with a focus on uncertainty modelling and quality control for spatial data, object extraction and change detection from satellite images and laser scanning (LiDAR) data, three-dimensional and dynamic modelling, and spatial analysis in GIS. He proposed and developed the principles of modelling uncertainties in spatial data and spatial analyses, and published the first book on this topic. Four major theoretical breakthroughs have been made through his research: (1) from determinant- to uncertainty-based representation of geographic objects in GIS, (2) from uncertainty modelling for static spatial data to dynamic spatial analyses, (3) from uncertainty modelling for spatial data to spatial models, and (4) from error description of spatial data to spatial data quality control. Because of these achievements, he received the Wang Zhizhuo Award, given by the International Society for Photogrammetry and Remote Sensing (ISPRS) in 2012, and also the State Natural Science Award by the State Council, China, in 2007. He served as president of the Commission on Theory and Concepts of Spatial Information Science, International Society for Photogrammetry and Remote Sensing (2008–2012), and president for the Hong Kong Geographic Information System Association (2001–2003). He also serves as an editorial board member for a number of international journals, including *IJGIS*, the top journal in GIS. He has published more than 130 *SCI* journal articles and more than 10 books.

Bo Wu is an associate professor with the Department of Land Surveying and Geo-Informatics of the Hong Kong Polytechnic University, Hung Hom. His current research interests are in the areas of digital photogrammetry and geographic information science (GIS), with a focus on image matching, robotic vision, fusion of imagery and laser scanning data, planetary mapping and planetary science, three-dimensional GIS, and applications. He worked on National Aeronautics and Space Administration (NASA)–funded projects on Mars and lunar exploration missions and the Chinese Chang'E-3 lunar exploration mission, and made useful contributions to the success of these missions. He served as the vice president of the Hong Kong Geographic Information System Association (2011–2013), and currently is serving as the co-chair of Working Group II/6 (Geo-Visualization and Virtual Reality) of the International Society for Photogrammetry and Remote Sensing (ISPRS). He has received many awards during his career, including the John I. Davidson President's Award for Practical Papers from the American Society for Photogrammetry and Remote Sensing (2014), the Duane C. Brown Senior Award from the Ohio State University, Columbus (2009), and the Nomination Award for the Top 100 National Excellent Doctoral Dissertation of China (2008).

Alfred Stein is a professor in spatial statistics and image analysis with the University of Twente, Enschede, the Netherlands. His research interests focus on statistical aspects of spatial and spatiotemporal data, such as monitoring data, in the widest sense. It includes optimal sampling, image analysis, spatial statistics, and use of prior information, as well as issues of data quality, fuzzy techniques, and random sets, all in a Bayesian setting. He is a member of the Research School for Socio-Economic and Natural Sciences of the Environment (SENSE). Since 2011 he has been the editor-in-chief of *Spatial Statistics*, the new leading platform in the field of spatial statistics. He is associate editor of the *International Journal of Applied Geoinformation and Earth Observation*, *Statistica Neerlandica*, and *Environmental and Ecological Statistics*. He has supervised more than 30 PhD students, and 12 PhD students are working under his supervision.

Contributors

F. Amanzougarene
EIVP: École des ingénieurs de la ville de Paris
Paris France

and

PRISM
Université de Versailles–SQ
Versailles, France

M. Beare
Beare Essentials Ltd.
Thurston, United Kingdom

Y. Bédard
Department of Geomatic Science and
Center for Research in Geomatics
Laval University
Québec City, Québec, Canada

M. Chachoua
EIVP: École des ingénieurs de la ville de Paris
Paris, France

N. Chrisman
Geospatial Sciences
RMIT University
Melbourne, Australia

Yongwan Chun
Geospatial Information Science
School of Economic, Political, and Policy Sciences
University of Texas at Dallas
Dallas, Texas

Chen-Yan Dai
College of Geographical Science
Fujian Normal University
Fuzhou, Fujian, People's Republic of China

M. R. Delavar
Center of Excellence in Geomatics Engineering in Disaster Management
Department of Surveying and Geomatics Engineering
College of Engineering
University of Tehran
Tehran, Iran

L. Diamond
Ordnance Survey
Southampton, United Kingdom

Andrew U. Frank
Department of Geodesy and Geoinformation
Vienna University of Technology
Vienna, Austria

J.-F. Girres
Université Paul Valéry Montpellier 3
Montpellier, France

and

Laboratoire COGIT
Université Paris–Est
Saint-Mandé, France

J. Goodwin
Ordnance Survey
Southampton, United Kingdom

Daniel A. Griffith
Geospatial Information Science
School of Economic, Political, and Policy Sciences
University of Texas at Dallas
Dallas, Texas

J. Grira
Department of Geomatic Science and
Center for Research in Geomatics
Laval University
Québec City, Québec, Canada

E. Guilbert
Department of Geomatic Sciences
Laval University
Quebec, Canada

Xincheng Guo
Chang'an University
Xi'an, People's Republic of China

A. Hackeloeer
BMW Forschung und Technik GmbH
Munich, Germany

J. Harding
Ordnance Survey
Southampton, United Kingdom

G. Hart
Ordnance Survey
Southampton, United Kingdom

D. Holland
Ordnance Survey
Southampton, United Kingdom

J.-H. Hong
Department of Geomatics
National Cheng Kung University
Tainan City, Taiwan

A. Hopfstock
Federal Agency for Cartography and Geodesy (BKG)
Frankfurt, Germany

M.-L. Huang
Department of Geomatics
National Cheng Kung University
Tainan City, Taiwan

Muhammad Imran
Geo-Information Science and Earth Observation (ITC)
University of Twente
Enschede, the Netherlands

A. Jakobsson
SDI Services
National Land Survey of Finland
Helsinki, Finland

F. Khamespanah
Department of Surveying and Geomatics Engineering
College of Engineering
University of Tehran
Tehran, Iran

K. Klasing
BMW Forschung und Technik GmbH
Munich, Germany

J. M. Krisp
Department of Geography
Universität Augsburg
Augsburg, Germany

Zhi-Lin Li
Department of Land Surveying and Geo-Informatics
Hong Kong Polytechnic University
Hung Hom, Kowloon, Hong Kong

and

Faculty of Geosciences and Environmental Engineering
Southwest Jiao Tong University
Chengdu, Sichuan, People's Republic of China

Eryong Liu
School of Resources and Environment
Fujian Agriculture and Forestry University
Fuzhou, Fujian, People's Republic of China

Zhiyong Lv
School of Computer Science and Engineering
Xi'an University of Technology
Xi'an, People's Republic of China

Milad Mahour
Geo-Information Science and Earth Observation (ITC)
University of Twente
Enschede, the Netherlands

L. Meng
Department of Cartography
Technische Universität München
Munich, Germany

Mingjing Miao
School of Printing and Packaging
Wuhan University
Wuhan, People's Republic of China

Gerhard Navratil
Department of Geodesy and Geoinformation
Vienna University of Technology
Vienna, Austria

R. Patrucco
Ordnance Survey
Southampton, United Kingdom

M. Pendlington
Ordnance Survey
Southampton, United Kingdom

A. Radburn
Ordnance Survey
Southampton, United Kingdom

S. Roche
Department of Geomatic Science
and
Center for Research in Geomatics
Laval University
Québec City, Québec, Canada

Wenzhong Shi
Department of Land Surveying and Geo-Informatics
Hong Kong Polytechnic University
Hung Hom, Hong Kong

Alfred Stein
Geo-Information Science and Earth Observation (ITC)
University of Twente
Enschede, the Netherlands

Hai Su
School of Printing and Packaging
Wuhan University
Wuhan, People's Republic of China

Bin Wang
Department of Land Surveying and Geo-Informatics
Hong Kong Polytechnic University
Hung Hom, Hong Kong

Pei Wang
First Institute of Photogrammetry and Remotes Sensing
Xi'an, People's Republic of China

David W. Wong
Department of Geography
University of Hong Kong
Pokfulam, Hong Kong

and

Department of Geography and Geoinformation Science
George Mason University
Fairfax, Virginia

Yaohua Yi
School of Printing and Packaging
Wuhan University
Wuhan, People's Republic of China

Ming Yu
College of Geographical Science
Fujian Normal University
Fuzhou, Fujian, People's Republic of China

and

Department of Land Surveying and Geo-Informatics
Hong Kong Polytechnic University
Hung Hom, Kowloon, Hong Kong

Yuan Yuan
School of Printing and Packaging
Wuhan University
Wuhan, People's Republic of China

M. Zare
Department of Surveying and Geomatics Engineering
College of Engineering
University of Tehran
and
Seismology Research Center
International Institute of Earthquake Engineering and Seismology (IIEES)
Tehran, Iran

K. Zeitouni
PRISM
Université de Versailles–SQ
Versailles, France

L. Zhang
Key Laboratory of Virtual Geographic Environment
Ministry of Education
Nanjing Normal University
and
Jiangsu Center for Collaborative Innovation in Geographical Information Resource Development and Application
Nanjing, People's Republic of China

Libin Zhao
First Institute of Photogrammetry and Remotes Sensing
Xi'an, People's Republic of China

Section I

Uncertainty Modelling and Quality Control

1

Uncertainty-Related Research Issues in Spatial Analysis

Daniel A. Griffith,[1] David W. Wong,[2] and Yongwan Chun[1]

[1]*Geospatial Information Science, School of Economic, Political, and Policy Sciences, University of Texas at Dallas, Dallas, Texas*

[2]*Department of Geography, University of Hong Kong, Pokfulam, Hong Kong, and Department of Geography and Geoinformation Science, George Mason University, Fairfax, Virginia*

CONTENTS

ABSTRACT This chapter enumerates different sources of error in spatial data. It highlights some recent cartographic developments in handling sampling error. We argue that uncertainty in spatial data needs to be addressed because error in spatial data affects the truthfulness of spatial patterns revealed by maps and the effectiveness of developing spatial models. We mention difficulties in defining neighborhoods and addressing aggregation error. Finally, we call for the inclusion of data quality information for each geographical feature in a spatial database as a prerequisite to addressing spatial data quality issues.

1.1 Introduction

Spatial data comprise two components: attribute and location information. Attribute information describes the nonlocational characteristics of features, whereas locational information indicates relative or absolute positioning of these features. Like aspatial data, attribute data have parameters that can be estimated with samples. Thus, one major source of uncertainty is sampling error (i.e., deviations of sample statistics from their corresponding population parameter values). Most often, the scoring of attributes also contains measurement error (i.e., differences between pairs of true and measured values). This additional major source of uncertainty involves proximity of an instrument reading to its corresponding real-world value and includes rounding of numbers and sometimes recording mistakes. Because all models are simplified descriptions of reality—Box (1979, p. 2) said, "All models are wrong, but some are useful"—these descriptions contain specification error (i.e., differences between reality and a model's representation of it); one goal of science is to minimize this error so that it is not too serious. Only approximate, rather than absolute, positions of features can be tagged to a coordinate system, resulting in location data also having error (i.e., deviations between true and approximate positions), introducing a fourth major source of error for georeferenced data. All four of these sources of error interact, impacting the quality of spatial data and spatial analyses, including mapping. In addition, features or geographical units can be merged or aggregated, perhaps due to confidentiality, data management, or representational concerns. Spatial aggregation can exacerbate and propagate error in spatial data from all four of these sources. Maps containing these errors are more uncertain—their representation of reality may be unnecessarily flawed.

1.2 Visualization of Uncertainty

Uncertainty in spatial data is a well-researched topic. When the U.S. National Center for Geographic Information and Analysis (NCGIA) formulated its research agenda about this topic in 1988, two initiatives were related to spatial data quality: accuracy of spatial databases and visualizing the quality of spatial information. The first initiative emphasized the sources and management of spatial data uncertainty (Goodchild and Gopal, 1989), whereas the second initiative focused on the visualization of spatial data error (Beard et al., 1991).

Researchers addressing the second NCGIA theme proposed many mapping designs and visualization methods to map data quality information (e.g., Kardos et al., 2005; Leitner and Buttenfield, 2000; MacEachren et al., 2005). Although many aspects of the uncertainty of given recorded observed values stemming from sampling, measurement, and locational and specification error can be represented reasonably well with existing methods, few of these methods have been adopted as standard practice or implemented in popular geographic information science (GIS) software. Reasons for including error information on maps have not been obvious to most users or readers. Map readers often look for geographical patterns, which emerge when the geographical distribution of a variable exhibits a systematic spatial arrangement. But, if observed values for a variable across locations are not statistically different, then the recognized pattern may not be true. Therefore, probably the most pragmatic reason to account for data error when mapping is to be able to determine whether differences between recorded observed values are significant.

Previous methods to map error fall short of assisting readers to differentiate between or compare recorded observed values. The *Atlas of United States Mortality* (Pickle et al., 1996) initiated an attempt to help readers discern attribute uncertainty across locations by including error bars. Sun and Wong (2010) evaluated different approaches to incorporate data reliability information in a choropleth map in the context of mapping estimates from the U.S. American Community Survey (ACS). Adopting the bivariate legend as the preferred design to incorporate error information in choropleth maps, they developed an ArcGIS extension that automates some of the steps in creating bivariate legend choropleth maps using ACS data, or any data, accompanied by the margin of error (MOE) information (Wong and Sun, 2013). The extension also includes several functions for map readers to compare values on a map, determining whether different reported observed map values are statistically different. Such functions are useful to help discern whether patterns that emerge from a set of reported observed map values are artifacts of errors. The extension is freely available to the public. However, more complicated comparisons of recorded observed values—ones that account for different types of error—are still difficult.

Recently, a new classification method to determine class breaks in choropleth maps was proposed (Sun et al., 2014). The class separability method considers error of the estimates in determining classes such that estimates in different classes are statistically different to a large extent. While this classification method may produce maps with relatively unbalanced classes, the spatial patterns revealed by the classification method are more reliable than those using other traditional classification methods, such as quantiles.

1.3 Impacts of Uncertainty on Spatial Patterns and Models

Although map readers search for spatial patterns on a map of observed values, more quantitatively oriented analysts prefer to quantify spatial patterns, often in terms of the level of spatial autocorrelation. Using similar concepts, cluster detection techniques often are used to identify attribute hot spots and cold spots. These spatial pattern analysis techniques assume that spatial data are relatively accurate. If the magnitudes of errors in recorded observed values from the various sources are available, they should be considered in evaluating the presence of spatial autocorrelation or spatial clusters (e.g., constructed confidence intervals should reflect them). Unfortunately, existing practices in evaluating spatial patterns, using either global or local statistics, completely ignore the presence of errors in spatial data, or assume that the data are relatively accurate, such that decisions made based upon the data are not compromised. These practices and assumptions definitely are not warranted (Citro and Kalton, 2007).

After evaluating the spatial patterns of response variable outcomes or events, the next logical step in a spatial analysis is to describe a relationship using aspatial and spatial models. Similar to the evaluation of spatial patterns, these models assume trivial levels of errors in both their predicted values (i.e., specification error) and their explanatory variables (e.g., measurement error). If variables have substantial levels of errors, a modelling effort should include the error information, and specification of the models and their predictions should be adjusted. Kriging furnishes one example of how this should be done, with its prediction map accompanied by its prediction error map (uncertainty), as well as its nugget effect (measurement or specification error) estimate. Often spatial models are formulated in the context of explanatory variables that are neighborhood characteristics, some of which may be derived by summarizing individual attribute outcome characteristics within neighborhoods. The majority of studies to date use existing artificial areal units (e.g., pixels of a given size and census tracts or blocks) as proxies for neighborhoods. But individuals residing within these areal units may not necessarily be influenced by the neighborhood defined by their artificial boundaries. Neighborhoods may extend into nearby units. Thus, the neighborhood characteristics measured for artificial units fail to accurately and precisely reflect real-world neighborhood characteristics.

1.4 Areal Unit Definitions and Spatial Data Aggregation

Accessing the effects of neighborhood characteristics on various attribute variable outcomes has been one of the major themes in spatial research (e.g., Kawachi and Berkman, 2003). Many studies adopt a simplistic and pragmatic approach to select existing areal units, such as census tracts or block groups, as proxies for neighborhoods. Thus, individuals geocoded to a unit, regardless of their socioeconomic and demographic differences, and their dispersions of locations within the unit, are assumed to have the same neighborhood demarcated by an artificial unit boundary. Apparently, definitions of neighborhoods are extremely complicated, and such simplistic one-size-fits-all approaches to determine neighborhood boundaries for people residing within the same statistical enumeration units work quite poorly in demarcating individual neighborhoods (e.g., Coulton et al., 2001; Diez-Roux, 2003). Such general practice introduces error in defining neighborhoods for individuals. Unfortunately, often subsequent analyses use such potential erroneous neighborhood definitions to explain attribute variable outcomes. Frequently, physical and socioeconomic characteristics of neighborhoods are used as covariates to explain attribute variable outcomes of individuals within neighborhoods, but the extent to which the individuals are associated with characteristics of a poorly defined neighborhood is completely ignored. Studies adopting the standard frameworks in neighborhood analysis in many societal aspects, such as public health outcomes and crime, potentially commit neighborhood assignment errors, failing to capture the neighborhood information accurately, and thus potentially produce misleading results and decisions.

The modifiable areal unit problem (MAUP), which refers to the inconsistency of analytical results arising from using spatial data tabulated for different resolution levels (scale effect) and for different zonal configurations of similar unit numbers (zoning effect) (Wong, 2009), is a recurring concern in the first theme of the NCGIA initiative (accuracy of spatial databases). Although no general solutions to the MAUP exist, such inconsistencies in results may be regarded as uncertainty associated with spatial data and have significant implications for and impacts on spatial studies (e.g., Grubesic and Matisziw, 2006; Wieczorek et al., 2012). Of the two subproblems constituting the MAUP, the zoning problem may be partially handled with spatial interpolation (Fisher and Langford, 1995), whereas the scale problem seems to defy any meaningful correction or adjustment. Luo et al. (2010) examine the impact of spatial aggregation on a statistical analysis of late-stage breast cancer. They show that aggregation changes data distributions of variables, and subsequently, statistical results are sensitive to a spatial aggregation level.

Griffith et al. (1998) conducted analyses at the individual, census block, census block group, and census tract levels and report that regardless of geographic resolution, pediatric blood lead levels contain positive spatial autocorrelation, with the same conspicuous spatial patterns and with selected covariate relationships that are stable across geographic resolutions. Griffith et al. (2007) discuss the properties of positional error in a street address matching exercise and the allocation of point locations to census geography units. Wong (2009) has written about the MAUP in general, its impacts on multivariate statistical analysis (Fotheringham and Wong, 1991), measuring segregation (Wong, 1997; Wong et al., 1999), and clustering patterns (Wieczorek et al., 2012). He also has introduced a correlation method that is less sensitive to the scale effect (Wong, 2001). Although many researchers treat aggregation error as one aspect of spatial data quality, most work on visualizing spatial data quality focuses only on sampling error, ignoring aggregation error (e.g., Leitner and Buttenfield, 2000; MacEachren et al., 2005; Pickle et al., 1996; Sun and Wong, 2010). Therefore, documenting and visualizing aggregation error from spatial data for multiple geographical resolutions are desperately needed.

In general, finer-resolution data (i.e., lower-aggregation level) use smaller areal units, and therefore have less of a scale effect and vice versa. Therefore, data tabulated for smaller areal units are preferable because they have less aggregation error. However, data for smaller areal units frequently are based on smaller samples, and therefore their sampling error levels become substantially greater. Aggregating smaller areal units to form larger units tends to yield more stable recorded observed values, such as disease rates. Bolstering such recorded observed values by borrowing information from nearby areal units also tends to stabilize them, but at a cost of additional error propagation or lowering the spatial resolution of the data. Thus, the role of aggregation error and its relationship with sampling error need to be investigated more thoroughly when highly aggregated data are the ones often used in practice.

1.5 Spatial Metadata for Data Quality

Wong and Wu (1996) introduce the concept of spatial metadata, data about the quality of each attribute for each feature in a GIS context. Recently, this spatial metadata concept has been adopted for such purposes as dissemination of the U.S. ACS and the U.S. Current Population Survey (CPS) data, where the distributed databases include the MOE for each estimate. However, such data quality information is limited to sampling error. The feature-based spatial metadata notion has been adopted by several prototype systems in managing feature-based metadata (e.g., Devillers et al., 2005; Gan and Shi, 2002; Qiu and Hunter, 2002). These systems are relatively complicated and deal with most data quality aspects listed in the U.S. Federal Geographic Data Committee (FGDC) metadata standards (FGDC, 1998). Systems with capabilities to manage and document spatial aggregation error, especially when the spatial data sets are for larger geographical entities comprising smaller areal units, essentially do not exist, with the exception of the Data Uncertainty Engine (DUE) (Brown and Heuvelink, 2007). DUE data quality information is derived through simulation experiments based upon assumptions about certain error sources and properties. For example, the DUE focuses on measurement and locational error and all but ignores sampling and specification error. Also, these types of systems are very sophisticated and involve complicated operations that often are beyond the comprehension of most data users, cartographers, GIS professionals, and some researchers.

Sometimes spatial data sets include uncertainty indices to accompany reported observed map values. For example, the ACS and the CPS include MOEs with their recorded values. Therefore, sampling error information for these georeferenced data is readily available for use in GIS work (Wong and Sun, 2013). When data at multiple geographical levels are available, some form of aggregation error can be derived empirically. Users at all levels also need to be encouraged to consider errors contained in spatial data in a routine and serious fashion, because the ramifications of not considering errors may be significant (Citro and Kalton, 2007). As pointed out by Heuvelink and Burrough (2002, p. 111), "[it] is crucially important to know how accurate the data contained in spatial databases really are, because without that knowledge we cannot assess the true value of the derived information, nor the correctness of the decisions it supports." Unfortunately, existing environments for handling spatial data are ill-equipped with tools to derive and manage aggregation error information, and the situation needs to be improved.

1.6 Conclusions

In conclusion, error in spatial data originates from multiple sources, and different types of error often tend to interact. Although disentangling these various interactions may be nearly impossible, if not impossible, understanding the nature of different sources of error and their interactions is the first step to advance this line of research. Spatial analysts need to continue to develop more sophisticated modelling procedures and tools to facilitate the handling of error during the use of spatial data, whether within a GIS, through mapping, or during spatial analysis. In parallel, these researchers need to continue to educate both the academic and general public, as well as all categories of data users, about "all [spatial] data hav[ing] error." And, they need to pursue a research agenda that addresses the visualization of uncertainty, impacts of uncertainty on spatial patterns and models, relationships between areal unit definitions and error propagation via spatial data aggregation, and general data quality, particularly in terms of metadata.

Acknowledgment

Research reported in this chapter was supported by the Eunice Kennedy Shriver National Institute of Child Health and Human Development of the National Institutes of Health under award R01HD076020. The content is solely the responsibility of the authors and does not necessarily represent the official views of the National Institutes of Health.

References

Beard, M.K., Buttenfield, B.P., and Clapham, S.B. NCGIA research initiative 7: Visualization of spatial data quality. NCGIA Technical Paper 91-26. University of California, Santa Barbara National Center for Geographic Information and Analysis, 1991.

Box, G. Robustness in the strategy of scientific model building. In R. Launer and G. Wilderson (eds.), *Robustness in Statistics: Proceedings of a Workshop.* Academic Press, New York, 1979, 201–236.

Brown, J.D., and Heuvelink, G. The Data Uncertainty Engine (DUE): A software tool for assessing and simulating uncertain environmental variables. *Computers and Geosciences,* 33(2), 172–190, 2007.

Citro, C.F., and Kalton, G. (eds.). *Using the American Community Survey: Benefits and Challenges.* National Academies Press, Washington, DC, 2007.

Coulton, C.J., Korbin, J., Chan, T., and Su, M. Mapping residents' perceptions of neighborhood boundaries: Methodological note. *American Journal of Community Psychology,* 29, 371–383, 2001.

Devillers, R., Bédard, Y., and Jeansoulin, R. Multidimensional management of geospatial data quality information for its dynamic use within GIS. *Photogrammetric Engineering and Remote Sensing,* 71, 205–15, 2005.

Diez-Roux, A.V. The examination of neighborhood effects on health: Conceptual and methodological issues related to the presence of multiple levels of organization. In I. Kawachi and L.F. Berkman (eds.), *Neighborhoods and Health.* Oxford University Press, Oxford, 2003, 45–64.

Federal Geographic Data Committee (FGDC). Content standard for digital geospatial metadata. FGDC-STD-001-1998. FGDC, Washington, DC, USA, 1998.

Fisher, P.F., and Langford, M. Modeling the errors in areal interpolation between zonal systems by Monte Carlo simulation. *Environment and Planning A,* 27(2), 211–224, 1995.

Fotheringham, A.S., and Wong, D. The modifiable areal unit problem in multivariate statistical analysis. *Environment and Planning A,* 23, 1025–1044, 1991.

Gan, E., and Shi, W. Error metadata management system. In W. Shi, P.F. Fisher, and M.F. Goodchild (eds.), *Spatial Data Quality.* Taylor & Francis, London, 2002, 251–66.

Goodchild, M.F., and Gopal, S. (eds.). *Accuracy of Spatial Databases.* Taylor & Francis, London, 1989.

Griffith, D., Doyle, P., Wheeler, D., and Johnson, D. A tale of two swaths: Urban childhood blood lead levels across Syracuse, NY. *Annals of the Association of American Geographers,* 88, 640–665, 1998.

Griffith, D., Millones, M., Vincent, M., Johnson, D., and Hunt, A. Impacts of positional error on spatial regression analysis: A case study of address locations in Syracuse, NY. *Transactions in GIS,* 11, 655–679, 2007.

Grubesic, T.H., and Matisziw, T.C. On the use of ZIP codes and ZIP code tabulation areas (ZCTAs) for the spatial analysis of epidemiological data. *International Journal of Health Geographics,* 5, 58, 2006.

Heuvelink, G.B.M., and Burrough, P.A. Developments in statistical approaches to spatial uncertainty and its propagation. *International Journal of Geographical Information Science*, 16(2), 111–113, 2002.

Kardos, J., Benwell, G.L., and Moore, A. The visualization of uncertainty for spatially referenced census data using hierarchical tessellations. *Transactions in GIS*, 9(1), 19–34, 2005.

Kawachi, I., and Berkman, L.F. *Neighborhoods and Health*. Oxford University Press, Oxford, 2003.

Leitner, M., and Buttenfield, B.P. Guidelines for the display of attribute certainty. *Cartography and Geographic Information Science*, 27(1), 3–14, 2000.

Luo, L., McLafferty, S., and Wang, F. Analyzing spatial aggregation error in statistical models of late-stage cancer risk: A Monte Carlo simulation approach. *International Journal of Health Geographics*, 9, 51, 2010.

MacEachren, A.M., Robinson, A., Hopper, S., Gardner, S., Murray, R., Gahegan, M., and Hetzler, E. Visualizing geospatial information uncertainty: What we know and what we need to know. *Cartography and Geographic Information Science*, 32(3), 139–160, 2005.

Pickle, L.W., Mungiole, M., Jones, G.K., and White, A.A. *Atlas of United States Mortality*. National Center for Health Statistics, Hyattsville, MD, 1996.

Qiu, J., and Hunter, G.J. A GIS with the capacity for managing data quality information. In W. Shi, P.F. Fisher, and M.F. Goodchild (eds.), *Spatial Data Quality*. Taylor & Francis, London, 2002, 230–250.

Sun, M., and Wong, D.W.S. Incorporating data quality information in mapping the American Community Survey data. *Cartography and Geographic Information Science*, 37(4), 285–300, 2010.

Sun, M., Wong, D.W., and Kronenfeld, B.J. A classification method for choropleth maps incorporating data reliability information. *Professional Geographer*, 2014. DOI: 10.1080/00330124.2014.888627.

Wieczorek, W.F., Delmerico, A.M., Rogerson, P.A., and Wong, D.W.S. Clusters in irregular areas and lattices. *Wiley Interdisciplinary Reviews (WIREs): Computational Statistics*, 4(1), 67–74, 2012.

Wong, D. The modifiable areal unit problem (MAUP). In A.S. Fotheringham and P.A. Rogerson (eds.), *The SAGE Handbook of Spatial Analysis*. Sage, London, 2009, 105–123.

Wong, D.W., and Sun, M. Handling data quality information of survey data in GIS: A case of using the American Community Survey data. *Spatial Demography*, 1(1), 3–16, 2013.

Wong, D.W.S. Spatial dependency of segregation indices. *Canadian Geographer*, 41(2), 128–136, 1997.

Wong, D.W.S. Location-specific cumulative distribution function (LSCDF): An alternative to spatial correlation analysis. *Geographical Analysis*, 33(1), 76–93, 2001.

Wong, D.W.S., Lasus, H., and Falk, R.F. Exploring the variability of segregation index D with scale and zonal systems: An analysis of thirty U.S. cities. *Environment and Planning A*, 31, 507–522, 1999.

Wong, D.W.S., and Wu, C.V. Spatial metadata and GIS for decision support. In J.F. Nunamaker Jr. and R.H. Sprague Jr. (eds.), *Proceedings of the 29th Hawaii International Conference on System Science (HICSS)*, Vol. 3: *Collaboration Systems and Technology*. IEEE Computer Society Press, Washington, DC, 1996, 557–566.

2

Spatial Statistical Solutions in SDQ to Answer Agricultural Demands Based on Satellite Observations

Alfred Stein, Muhammad Imran, and Milad Mahour

Geo-Information Science and Earth Observation (ITC), University of Twente, Enschede, The Netherlands

CONTENTS

ABSTRACT Satellite observations are addressing agricultural demands at various scales: from the plant level to the national level. Such observations, however, have their specific accuracy, in terms of both their attribute value and their location coordinates. Moreover, their intended use is different. In this chapter, we study the role of satellite images at the two extreme cases. We first address mapping at the field scale, where subpixel resolution mapping is done to identify individual plants. Second, yield mapping is performed at the national level where upscaling is done using geographically weighted regression. The study shows how at these extreme cases, skillful spatial statistical methods collect information that may be useful for decision making, and what the role of spatial data quality is. It concludes that the quality of the spatial information is to be further improved in order to have even better statements than to date.

2.1 Introduction

In a broad range of agricultural studies, the use of spatial data is important. The main reason is that spatial variation is present at various scales in space and in time (Finke and Stein, 1994). Such variation affects decision making by the farmer as the user of this information: it may lead to various dosages and other treatments at finer- and coarser-scale levels to be applied at various moments (Bouma, 1997). Spatially, variation occurs at the within-parcel level that we consider in this study as the finest-scale level. It extends toward the between-parcel level that is similar to the within-farm level, toward the between-farm level, and even beyond. There is a clear scale issue at stake; for example, we often consider a parcel as a single cropped field, although we realize that in modern farming systems, mixed cropping also occurs at the within-parcel scale, and hence the within-field variation occurs at the plant level or for groups of plants. The precision of such quantified variation is critical; it should lead to better applications, and not necessarily to applications that are overly refined. In all, the quality of the data, relating their precision to fitness for use, is therefore most important.

Measuring data relevant for agricultural purposes can be done in many ways. In this chapter, we consider field measurements in relation to remotely sensed data. First, relevant field measurements can be taken, for example, on vegetation, weather, soil, and groundwater. Apart from the type of variable itself, these also establish a measurement frequency in time. For example, a soil map is usually established once and is for several years considered to be fixed, whereas vegetation changes during the growing season, and weather changes can be even more frequent. Second, remotely sensed data cover many fields and farms and are collected at an increasingly high frequency: with the advent of unmanned aerial vehicles, timely information can be collected at high frequencies, although requiring skillful processing, whereas satellite images are collected at fixed intervals covering larger areas. The aim of most of the collected data is in the farmer's decision making, possibly in relation to decision making at the higher (governmental) levels.

Use of both field data and remote sensing data requires zooming toward the scale of application. For estimates of large areas, like a region or a country, zooming out is essential. Field data are collected at, for example, small farms or individual soil pits, which have to be extrapolated toward the larger areas. This induces various forms of uncertainty, like varying natural conditions, varying management conditions, and also varying climate conditions. For such processes, remote sensing data serve a proper purpose, as these can relatively simply cover such areas. Alternatively, remote sensing images are also increasingly used for zooming *in*. The levels of zooming in become increasingly refined; these days satellite resolution is down to 0.5 m and less at the ground level. There is still a large amount of progress, in particular as spatial resolution becomes an almost negligible issue when using drones that are now affordable to all.

The quality of spatial data is connected with uncertainty, which in turn relates to accuracy and precision. Uncertainty is a wide concept that refers to lack of precision and lack of accuracy. Accuracy is an issue of the data mainly. Accurate data imply that in principle, the right variable can be measured with a high precision, that is, with many decimals, whereas inaccurate data mean that a relatively crude measuring device is being used. Tempting as it is, highly accurate data are not required for each and every application. In addition, uncertainty refers to lack of deterministic knowledge that could be captured in a process-based model and that in principle could be overcome by sharper definitions, sharp classes, and in that sense, more precise information. In the absence of any of these, fuzzy methods have been proposed. It depends on the fitness for use whether accurate data are required.

The aim of the current study is to show two ends of the scale spectrum in terms for the use of remotely sensed images. On the one hand, it shows how precise data can be obtained for farmers' use, whereas on the other side of the spectrum, we will upscale the data toward a whole country. We consider error propagation and illustrate the study with practical examples.

2.2 Variation of Agricultural Data

When collecting agricultural data, we distinguish between spatial variation and nonspatial variation. On top of this, there is variation in the time domain. The basis for each analysis of agricultural data is the concept of a probabilistic variable X; measurements taken on X are denoted by $x_1, \ldots, x_n$. Such data can be randomized, as a randomization will not change the basic descriptive statistics like the median, the mean, the variance, and percentiles. Referring to spatial data, we commonly include the spatial location s, making the variable $X(s)$ with observations $x(s_1), \ldots, x(s_n)$. These data cannot be randomized, as the spatial pattern reflects their values. If no changes occur in time, however, we may make multiple measurements in time. If this is not realistic in any sense, then the variable X should be related to both space S and time T, being denoted as $X(s, t)$, and at k observation times and n spatial locations with observations $x(s_i, t_j)$, $i = 1, \ldots, n$ and $j = 1, \ldots, T$. A simple example is the soil temperature: it can be considered nonspatial. For example, in a single parcel with no shadow, the variation is likely to be small, and one observation can be enough to characterize it, although it is preferred to make, say, five observations that are replicates of each other, leading to the mean value within a field. It can be considered a spatial variable, for example, if measured at approximately the same time, resulting in a map or displaying the maximum or minimum temperatures during a day; it can also be considered a spatiotemporal variable if variation occurs—and is relevant—in space and time.

A natural platform to integrate the different data is in a geographic information science (GIS). Since long, the use of geostatistics has been advocated to make a map out of the observations that shows the spatial variation. The use of geostatistics helps to quantify the spatial (and spatiotemporal) variation, create maps, and quantify the uncertainty. Spatial interpolation simulation procedures have been used in this respect. In space, geostatistics models the variation between locations separated by a distance h, denoted as s and $s + h$, by determining half their expectation. In the space–time domain, this is extended to the variations between locations s, t, and $(s + h_x, s + h_t)$, where h_x is the distance in space and h_t is the distance in time.

2.3 Error Propagation

Often one is not interested in the direct observations themselves but in a derived (indirect) observation. There is a model (function) between the data and the outcome. Such a model could be a simple function, a GIS operation, or a complicated deterministic model.

It is a fact of life, however, that data (maps) stored in the GIS database are rarely, if ever, error free. Causes are generalization, digitization, measurement, classification, and interpolation errors, whereas data are often generated using modelling. As a consequence, errors will propagate through GIS operation. A key research question is that given the errors in the inputs to the GIS operation, how large are the errors in the output?

Obviously, garbage in implies garbage out; hence, uncertainty in the observations leads to uncertainty in the results. For these reasons, many studies advocate the use of the mean of a limited set of observations as the input to a model. Realizing the inherent spatial and spatiotemporal variation, however, the appropriate scale level is essential. When a model is applied at the parcel scale, then average data values over that parcel serve as an input. For such data the within-parcel variation cannot be observed or analyzed any more. In a slightly more general setting, we may imagine that maps have been created as data layers and that these maps serve as an input to a model. We then obtain an output map Z as a function $g(.)$ of the input maps $A_1, \ldots, A_m$:

$$U = g(A_1, A_2, \ldots, A_m)$$

Examples include the following:

- An elevation map serves as an input to determine the slope map. Having access to a Digital Elevation Model (DEM), we can calculate the slope angle as *slope* = *g*(*elevation*), where *g*(.) is a spatial differential equation and $m = 1$. There is little uncertainty in the function *g*(.), the main source of uncertainty being the accuracy of the DEM.

- Using the Universal Soil Loss Equation (USLE), we can calculate the erosion risk, using an equation like *erosion risk* = *g*(*land use, slope, soil type*). Here, $m = 3$. A main source of uncertainty is the function *g*(.), and we may notice that weather patterns are not included as explanatory variables in *g*(.).
- A *yield map* = *g*(*yield observations*), where *g* is an interpolator. A geostatistical interpolator *g*(.), like ordinary kriging, relies on a set of assumptions, like stationarity, which may not hold, and on appropriate estimation of the variogram. Also, from the past we have noticed that the data include a change in accuracy, if the yield is determined continuously by an integrated observing and yielding device: if the yield is measured at a site that was already partly yielded by an earlier visit of the device, then a lower value is registered than what would be realistic.
- A *national yield map* = *g*(*yield data, other data*), where *g*(.) is a statistical model, such as geographically weighted regression model. Uncertainty is here due to the limited accuracy of the data and the validity of the statistical model.
- A *super resolution map* = *g*(*image*), where *g*(.) is a super resolution mapping (SRM) operator with the coarse resolution map as input and a classification system that is relevant at a finer resolution level. The function *g*(.) downscales the image to a finer level, proportional with a scale factor to the coarse level.

2.4 Error Propagation

Important in the functions is the propagation of the uncertainty, from the input uncertainty to uncertainty in the output. The most common way is error propagation. Error propagation analysis involves three steps: definition of an error model for quantitative spatial attributes, identification of the error model (estimation of its parameters), and performance of the actual error propagation analysis. Using a statistical approach, the main focus is on quantifying the method of measurements. Repeated measurements under similar circumstances may lead to different outcomes. This is due to random influences and can be quantified by the variance σ^2. The main problem then is, what are similar circumstances? An essential issue in error propagation concerns the difference between repeatability and reproducibility. *Repeatability* reflects the situation that measurements are carried out on identical objects by one observer with one technique, under identical conditions, shortly after each other, σ_r^2. *Reproducibility* reflects the situation that measurements are carried out on similar objects by different observers, by different techniques, at different conditions at larger time steps, σ_R^2. Commonly, we have that $\sigma_R^2 > \sigma_r^2$. Repeatability and reproducibility are two extremes, whereas intermediate combinations are possible as well. For an appropriate error analysis it should be clear how measurements have been carried out. A complicating factor is that identical objects may still show some variation—reflectance of one forest stand may be slightly different from reflectance of another stand, due to the different composition of the trees, even of the same density, for example, caused by different intertree distances, different soil and hydrological conditions, or different varieties.

As a simple example, we consider a function that is fully known and consider uncertainties in the input variables and for the moment do not consider uncertainty in the function itself. A standard GIS operation is the calculation of a polygon area from the coordinates of the vertices that describe the polygons' boundaries. Uncertainty in the coordinate values results in uncertainty in area estimates, resulting in uncertainty in the estimated utility value of the polygons.

The polygon area is calculated from the coordinates of its n vertices. The trapezoid rule requires that the vertices are sorted in a counterclockwise direction so that vertex $i - 1$ comes just before and vertex $i + 1$ just after vertex i. A polygon is described by a closed loop; hence, $s(0) \equiv s(n)$ and $s(n + 1) \equiv s(1)$. The polygon area $A(p)$ is then calculated as

$$A(p) = \frac{1}{2} \cdot \sum_{i=1}^{n} x(i) \cdot \left(y(i+1) - y(i-1)\right)$$

Using the statistical package R, we used three points $A = (0, 0)$, $B = (5, 0)$, and $C = (2.5, 4)$. Clearly, these three points lead to a triangle with an area equal to 10. When allowing inaccuracies to enter the coordinates, we may apply a value of, say, $\sigma = 0.1$ in both x and y coordinates and then compare these results with $\sigma = 1.0$ in both x and y coordinates and repeat this 100 times (Figure 2.1).

2.5 First Application: Zooming In—Super Resolution Mapping

The first application considers grape trees in Iran, where rows of plants and plants are characterized individually (Mahour, 2013). The objective was to estimate the actual and potential evapotranspiration (ET) at plant scale for

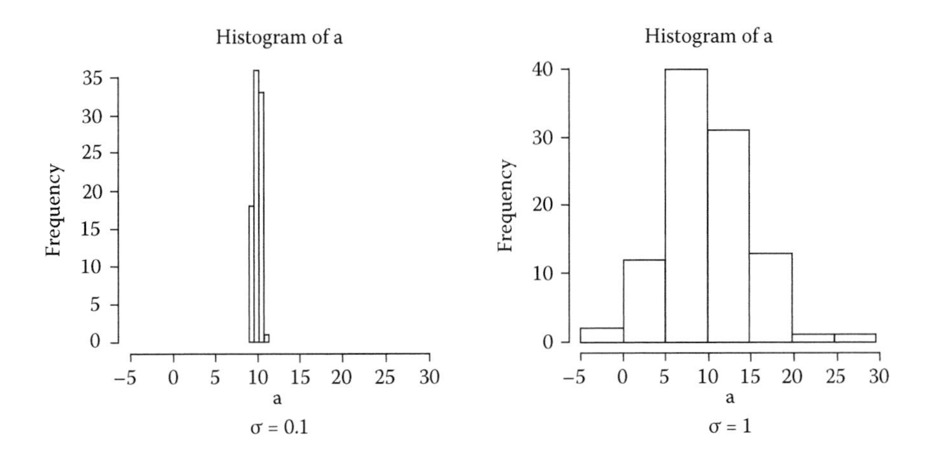

FIGURE 2.1
Error propagation to the area of a triangle, using an uncertainty of 0.1 unit in the coordinates (left) and 1 unit in the coordinates (right). Notice that for large uncertainties, negative areas were also obtained.

vineyards. Satellite information was available at different resolutions. One of the problems we faced was to supply reliable information on ET using the surface energy balance system (SEBS) model. A GeoEye satellite and an UltraCam image were used for identifying the plants, whereas a Landsat image was used for retrieving ET.

Precise information should be obtained from individual fields. The main question to address was whether we could use both spectral and spatial information of a high-resolution satellite image to detect *rows* and *individual plants*. Hence, the problem statement could be phrased as the development of a decision support system for water resource management in precision agriculture (PA). To solve the problem, we realized that in the absence of high-resolution multispectral data, we could use super resolution mapping (SRM). We combined this with the SEBS as a model to determine ET. In this we explored the possibility of assessing plant water deficit using SRM and application of SEBS. The data were obtained from a GeoEye-1 satellite image, an UltraCam digital aerial photo, a Landsat 5 TM satellite image, and meteorological data. Notice that only the Landsat 5 image could be used within the SEBS model, being the sole satellite with a temperature band. Meteorological data were available from a field station at the site to assess the quality of water requirements at the plant scale using SRM.

Super resolution mapping (Tolpekin and Stein, 2009) is applied to a land cover classification, where resolution of the output map is higher than that of the input image. A set of land cover classes $C_\alpha = 1, \ldots, L$ is defined, and pixels are defined that correspond to each class α. Pixel values are modeled with the normal distribution. Further, a training set is used to estimate the mean vector and covariance matrix C of the normal distribution for each C_α. We imagine the existence of a fine-resolution image A with pixels a_i and a classified image $a_i \rightarrow c_\alpha$, where $\alpha = 1, \ldots, L$. What is observed is a coarse-resolution image B with pixel values b_j that have a pixel area that is S^2 times larger than that for a_i. The goal of SRM is, given b_j, to estimate c_α. We applied the following assumptions: the pixel values follow the normal distribution, there is no spatial autocorrelation, and there is no effect from the point spread function.

Hence, a mixed pixel value b_j follows a multivariate normal distribution with mean μ_j and covariance matrix Σ_j being linear mixtures where $\theta_{\alpha j}$ is the fraction of class α in pixel b_j.

$$\mu_j = \sum_{\alpha=1}^{L} \theta_{\alpha j}\mu_\alpha \qquad \Sigma_j = \frac{1}{S^2}\sum_{\alpha=1}^{L} \Sigma_\alpha$$

The likelihood energy is modeled as the Mahalanobis distance between b_j and μ_j with covariance Σ_j:

$$U_{lik}\left(y|c\right) = \sum_{i=1}^{N} \frac{1}{2}\left(y_j - \mu_j\right)' \Sigma_j^{-1}\left(y_j - \mu_j\right)$$

The maximum a posteriori (MAP) solution of the problem corresponds to the minimum of the global energy with respect to the SRM c, where λ is a smoothness parameter:

$$U\left(c|y\right) = \lambda \cdot U_{prior}\left(c\right) + \left(1-\lambda\right) \cdot U_{lik}\left(y|c\right)$$

where $U_{prior}(c)$ is the prior energy. A MAP solution is obtained using simulated annealing:

$$P\left(c|y\right) \propto \exp\left[-\frac{U\left(c|y\right)}{T}\right]$$

where T is a temperature parameter that is decreased step-by-step during iterations. Class separability is a measure of the similarity between the classes. The divergence D and the transformed divergence TD for any pair of classes α and β with covariances equal to Σ_α and Σ_β, respectively, are

$$D_{\alpha\beta} = \tfrac{1}{2}(\mu_\alpha - \mu_\beta)'\left(\Sigma_\alpha^{-1} + \Sigma_\beta^{-1}\right)(\mu_\alpha - \mu_\beta) + \tfrac{1}{2}Tr(\Sigma_\alpha + \Sigma_\beta)(\Sigma_\alpha^{-1} + \Sigma_\beta^{-1})$$

$$TD_{\alpha\beta} = 2\left(1 - e^{-D_{\alpha\beta}/8}\right)$$

The smoothness parameter is an internal parameter of Markov Random Field (MRF)-based SRM. It needs to be estimated before application. Available estimation techniques aim to optimally describe training data, for example, a subset of a fine-resolution reference land cover map. We take $\lambda \le \lambda^*$, where

$$\lambda^* = \frac{1}{1 + \frac{2S^2}{D_{\alpha\beta}}\gamma}$$

Notice that if $\lambda > \lambda^*$, then the model will lead to oversmoothing. On the other hand, a too low value of λ does not fully exploit the prior information in the model. Therefore, $\lambda = \lambda^*$ is the optimal value. The solution relates λ^* to the scale factor S and the divergence D. We notice that the expression for λ^* includes a parameter γ ($0 < \gamma < 1$) that is related to the prior energy coefficients, the neighborhood window size, and a pixel class configuration in a specific image. Once γ is found, for example, estimated for a single combination of S and TD, the scaling behavior of λ for other values can be predicted.

We applied these methods first to individual rows of the plants that were identifiable from the image, and then continued to the individual plants. Figure 2.2 shows image fusion (c) between the actual ET image (b) and the Normalized Difference Vegetation Index (NDVI) image (a) using the Gramm–Schmidt fusion method. Figure 2.3 shows the obtained SRM results of a row and three individual plants for allocating actual ET.

The study showed that SRM is able to increase the resolution of classifications obtained from coarse-resolution input images for supporting precision agriculture. Image fusion of the SEBS results with SRM provided a meaningful end product for decision making in irrigation networks. Most important, however, the study showed that the level of zooming in at this stage reaches almost the level of detail of the individual plants. This is under the

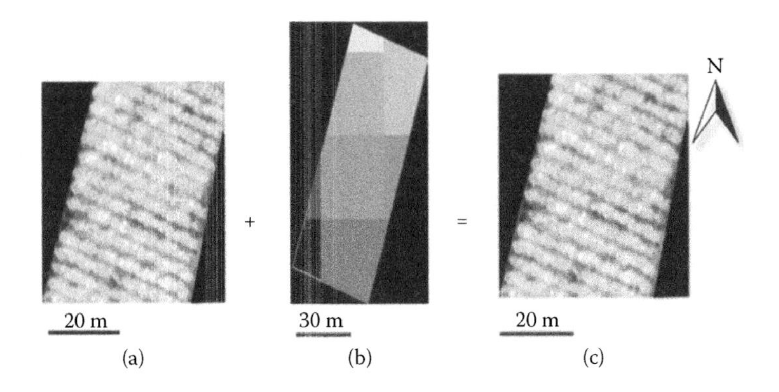

FIGURE 2.2
(a) High-resolution NDVI image. (b) Actual Evapotranspiration (AET) image obtained after applying the SEBS model. (c) Result after using Gramm–Schmidt image fusion.

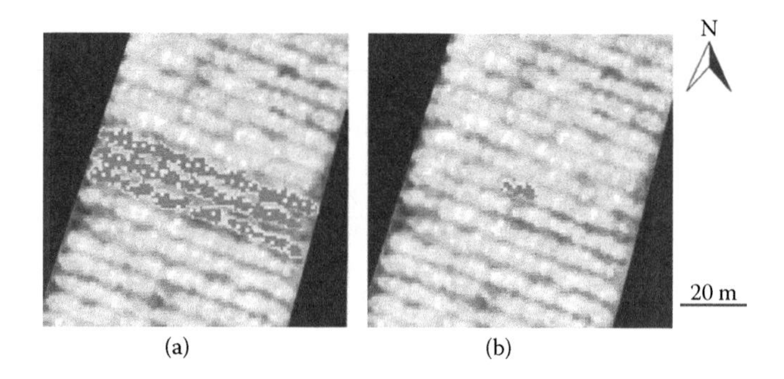

FIGURE 2.3 (See color insert.)
Combination of SRM with the actual ET (a) zooming in at three individual rows and (b) zooming in at three individual plants.

condition that a clear class distinction is feasible between the background and the plants. In a similar study on potatoes, for example, we observed that a close coverage of a row prohibits identification of the (apparently smaller) individual plants. Possibly, use of a Unmanned Aerial Vehicle (UAV) may yield more detailed results in the future (Figures 2.2 and 2.3).

2.6 Second Example: Zooming Out—Mapping Crop Yield in West African Rain-Fed Agriculture Using Global and Local Spatial Regression

This application concerns a wall-to-wall mapping of bioeconomic farming models. It is based on Imran et al. (2013). The reported crop yields in developing countries are often highly spatial variable due to varying agroecological conditions, soil types, and management factors. There could be explanatory variables: these are collected during ground-based field surveys for selected sites in various administrative units. Crop yield mapping addresses the following question: How should we model spatial variability of crop yields (and its explanatory variables) at the national scale? The study is applied to the country of Burkina Faso, with the aim to understand the spatial trends in agroclimatic zones.

Data may have different accuracies, as they can be collected from ground surveys, regional and national GIS products, or global remote sensing products. The fitness for use is defined here as the degree to which the data contribute to make the maps at the different scales. The data were collected as follows. Crop surveys were obtained from the national statistical data where crop yields were sampled from one representative terrority per district. The tropic-warm-semiarid and tropic-warm-subhumid agroecological zones were identified. Soil properties were obtained as regional and national GIS products. Available GIS layers were calcareousness (%), loam content (%), sand content (%), and water logging capacity (%). Also, topographic data like slope (°) and elevation (m) and the climate variable rainfall (mm) were available. The socioeconomic variable rural population density was provided as a proxy for labor availability. Remote sensing products (available at a global coverage) were collected as a time series of SPOT NDVI 1 km 10-day composites for the period covering the crop-growing season. These were analyzed using a principal component analysis to transform the correlated spectral and temporal image bands into principal components of the NDVI data.

Geographically weighted regression (GWR) was carried out to interpolate crop data from the field scale toward the national scale (Fotheringham et al., 2002). GWR is a location-specific spatial statistical technique. It is based on estimated parameters in regions with crop data and is then applied at regions where explanatory data are available but crop data are not.

$$Y(s) = \beta_0(s) + \sum_k \beta_k(s) X_k(s) + \varepsilon(s)$$

$$\hat{\beta}(s) = \left(X^T W(s) X\right)^{-1} X^T W(s) Y(s)$$

$$W_{ij}(s) = \exp\left(-\frac{1}{2}\left(\frac{d_{ij}}{a}\right)^2\right)$$

Here, $\beta_0(s)$ is the location-dependent intercept term, the $\beta_k(s)$ are the trend parameters according to the kth explanatory variable, also depending upon location s, of which the values are contained in the vector $X_k(s)$, and $\varepsilon(s)$ is the location-dependent error terms. The cap identifies the estimated parameters, X denotes the matrix consisting of the $X_k(s)$, the matrix $W(s)$ contains the weighting according to neighboring pixels, and $Y(s)$ are the observations. In order to facilitate estimation, Gaussian weighting was used as the kernel function, and hence the spatially weighted matrix $W(s)$ consists of the negative exponential of the squared distances d_{ij} between pixels that are scaled by a parameter a. For crop yield maps at the national scale, separate GWR models were established for each sampled location and local coefficients were estimated. GWR-based modelling could be used for mapping. The significantly contributing variables rainfall and topography were incorporated, as were the two agroclimatic zones of the country (Figure 2.4).

GWR models well explained the subregional yield variability. Interpolating of GWR local coefficients generated accurate yield maps. We see further prospects in solving spatial data quality issues, with the further development of the satellite launching programs. Expanding the methodology to several countries, a wall-to-wall mapping is achievable. The type of information is

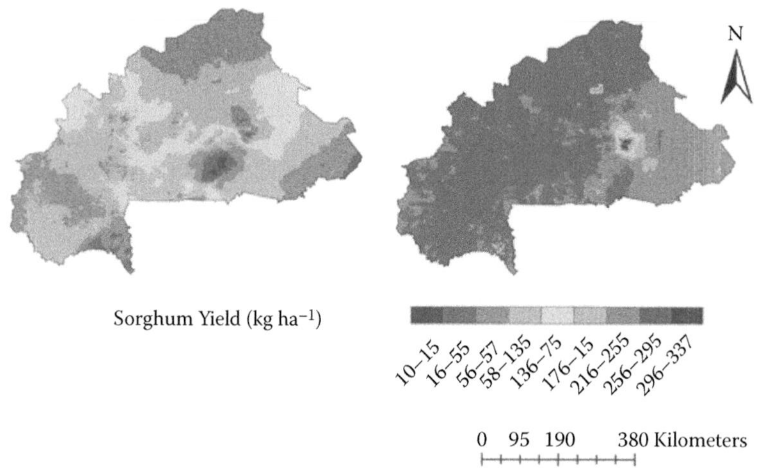

FIGURE 2.4 (See color insert.)
Predictions of sorghum yield at the national scale of Burkina Faso (left) and prediction errors (right). Both maps are able to show a large variation, due to the use of remote sensing images.

in particular important if food security is a major issue. In those circumstances, satellite images must be properly collected at those moments when the required information is needed most.

Spatial statistical tools are of help to identify spatial data quality. They show the locations of high and low spatial uncertainties and are in this sense able to deal with spatial variation.

2.7 Final Remarks

As the current study showed, remote sensing imagery can be used to assist in scaling up and scaling down spatial information. It is a fruitful source of important information. For downscaling (i.e., going to a finer resolution), the combination with SRM turned out to be most helpful and can be used to go almost to the plant level. The end of this line is not yet in sight, and we can assume that further progress will be made shortly. In particular, we see that increased spatial resolution and increased frequency in the time domain can help to overcome current limitations in spatial resolution. For upscaling, that is, going toward a coarser scale and hence toward larger areas of land, remote sensing images provide a wide and detailed coverage. They need not be taken as is but can be further investigated, for example, by taking averages, using derived products, and tuning classification so that they can be of the highest assistance.

In the future we see a better link of societal questions with the scientific approaches. For precision agriculture, we may imagine a closer relation with agricultural decision making that expands beyond issues of water availability and crop water requirement to also include alleviation of effects of pests and diseases at an earlier stage by identifying crop stress. Deepening the scientific approaches may lead to better statements, where issues of spatial uncertainty are increasingly addressed. Better understanding of what makes some spatial data of higher quality than other spatial data (not just root mean square (RMSE) or amount), and hence better understanding of the quality-generating concepts, may be of help as well. This might lead, for example, to a better understanding of the complex *semantics* of quality, increased attention to metadata quality, further exploration and analysis of uncertainty propagation beyond error propagation, and engagement of users of spatial data in Spatial Data Quality (SDQ).

References

Bouma, J. 1997. General reflections. In *Precision Agriculture: Spatial and Temporal Variability of Environmental Quality*. Ciba Foundation, John Wiley, Chichester, pp. 231–242.

Finke, P.A., and Stein, A. 1994. Application of disjunctive cokriging to compare fertilizer scenarios on a field scale. *Geoderma*, 62(1994), 247–263

Fotheringham, A., Brunsdon, C., and Charlton, M. 2002. *Geographically Weighted Regression: The Analysis of Spatially Varying Relationships*. John Wiley & Sons, New York.

Imran, M., Stein, A., and Zurita Milla, R. 2013. Modelling crop yield for West African rainfed agriculture using global and local spatial regression. *Agronomy Journal*, 105(4), 1177–1188.

Mahour, M. 2013. Exploring the application of SEBS for precision agriculture: Plant scale vineyard. MSc thesis, University of Twente.

Tolpekin, V., and Stein, A. 2009. Scale effects and parameter optimization for super resolution mapping from satellite images with Markov random fields. *IEEE Transactions in Geoscience and Remote Sensing* 47(9), 3283–3297.

3

First, Do No Harm: Eliminating Systematic Error in Analytical Results of GIS Applications

N. Chrisman[1] and J.-F. Girres[2,3]

[1]*Geospatial Sciences, RMIT University, Melbourne, Australia*

[2]*Université Paul Valéry Montpellier 3, Montpellier, France*

[3]*Laboratoire COGIT, Université Paris–Est, Saint-Mandé, France*

CONTENTS

ABSTRACT Geographic information science (GIS) applications compute analytical results composed of geometric measures such as perimeter, distance between objects, and area. Usual practice operates in Cartesian coordinates on map projections, inducing a systematic variation due to scale error. The magnitude of these errors is easy to foresee, though rarely corrected in analytical reports. Solutions to this error can be implemented through a number of alternative procedures. Professionals engaged in geospatial applications have an ethical obligation to minimize the avoidable distortions in reported results.

3.1 Introduction

3.1.1 Motivation

Research on spatial data quality considers many sources of variation in the data used for spatial analysis, usually with a stochastic approach. This chapter returns to a subject often neglected, though totally obvious: map projections. As we describe in all entry courses into the discipline, all map projections distort geometric properties in predictable, systematic ways (Snyder, 1987, p. 3). The current set of projections most used for geographic information science (GIS) applications were established as official coordinates. The purposes of these official adoptions privileged angles, and therefore nearly universally chose conformal projections. Angles were central to the data collection procedures of an earlier era. However, most of the analytical uses of GIS are more likely to report distances and areas as measures, though these are not preserved by conformal projections. Though the map-scale errors are well known and perfectly predictable, they remain untreated in reporting results. While the focus of data quality research has been on the statistical variability of features (see Griffith, 1989; Shi and Liu, 2000; Leung et al., 2004), we should consider all sources that make our results deviate from ground truth. Some errors are systematic, not random. Systematic errors are often easier to determine and thus to remove.

3.1.2 Aims and Overview

This chapter adopts the precautionary principle that any profession should consider: first, do no harm. The current practice of GIS typically moves directly into a map projection with only the most limited consideration of alternatives. This choice of projection may do harm to the application, and we demonstrate the magnitude of this effect in some typical use cases. Then we present some solutions to avoid this effect.

This chapter considers the intended variability in common projection systems and official coordinate systems, and then continues to the actual practice of these systems that extend frequently into zones beyond the design. The current practice of GIS applications involves study areas of much greater extent than the designers of official coordinate systems envisioned. Hence, study areas frequently extend over zone boundaries, and convenience of treatment leads to bending the rules.

This chapter builds on a recent PhD from France (Girres, 2012) on the error consequences for measures of perimeter and area. One section of this work dealt with projections, the subject of this chapter. These results demonstrate that the deviations are potentially significant. In addition, the error is strongly spatial so that certain regions have predictable under- and over-representation. Certain practices are advanced to mitigate these effects, but a radical solution must also be contemplated.

3.2 Error in Projections

3.2.1 Overview

The distortion introduced by a map projection is an analytical result known since Tissot (1881) or before (for full coverage, see Snyder, 1987, p. 20; Kimerling et al., 1995). For the purposes of this chapter, we will focus on two functions typically calculated with GIS data: perimeter and area. Both of these are related to the scale factor of the projection, a local parameter that measures the instantaneous variation in distances. It is possible to have no scale distortion at one point, or along one or two lines. These are the locus of tangency, where the projection surface is in contact with the reference ellipsoid (and ultimately the ground surface). Snyder (1988) has extended this to oblate footprints, but the principle remains. Figure 3.1 presents a Mercator projection in its normal aspect where the line of tangency is the equator. Scale is exaggerated away from the equator in order to maintain the property of conformality.

In certain cases, the scale factors in latitude and longitude balance out so that the area result is equivalent (or equal area). Equivalent projections distort distances, but in a manner that areas are preserved. However, many of the projections used routinely are designed to preserve angles (conformal) by having the same scale factor in all directions (isotropic).

3.2.2 Specific Characteristics of Conformal Projections Used in Reference Systems

One heritage of topographic mapping practices is the institution of official projections (called national grids, state plane coordinate systems, or some other title). This topic, while arcane, seems to capture inordinate interest due to the vast number of known spatial reference systems. The International Organization of Oil and Gas Producers (originally OGP) has curated a list of coordinate reference systems (OGP) that has grown to 4300 or more (known as the EPSG registry: http://www.epsg-registry.org/). Among these thousands, there is a lot of regularity. Examples below will reference the EPSG number for

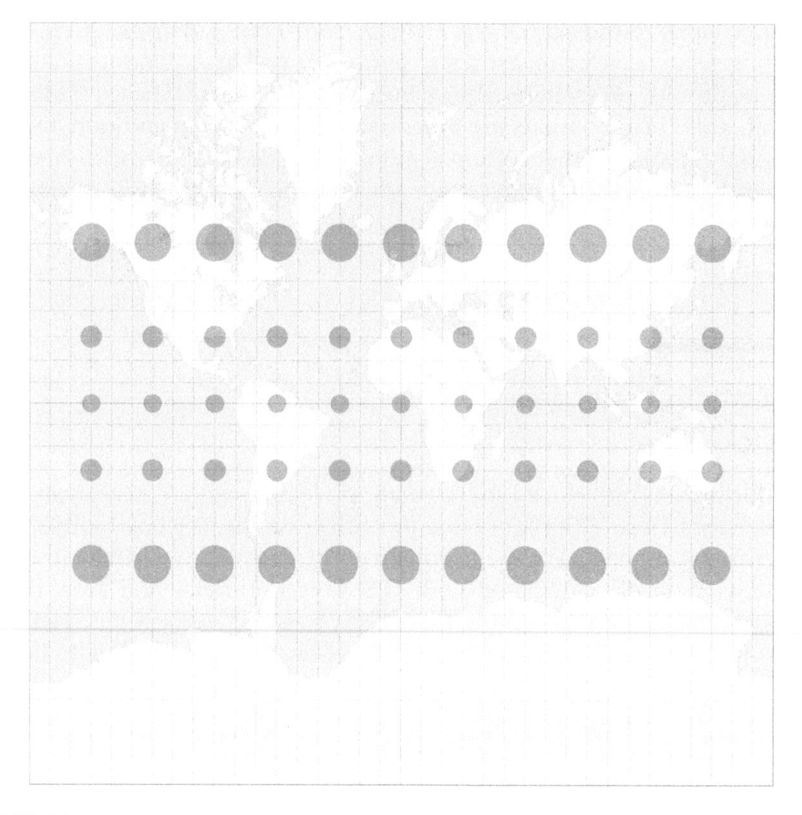

FIGURE 3.1
A conformal projection (Mercator, equatorial aspect) with exaggerated Tissot circles. (From Kuhn, S., http://commons.wikimedia.org/wiki/File:Tissot_mercator.png, Creative Commons attribution license.)

clarity. The most usual choice for official coordinates is the transverse Mercator (TM). In the version termed the Universal Transverse Mercator (UTM), developed for military mapping, the coordinate system applies worldwide with strips that are 6° of longitude wide. Some countries (including Australia) adopt UTM directly as their official coordinates (MGA, EPSG 28348-58), but many use smaller zones with tighter error properties (such as Modified Transverse Mercator (MTM) in Canada or Gauss–Kruger in Germany).

Certain other jurisdictions have adopted the Lambert conformal conic. Together with some oblique versions of the transverse Mercator, they cover the vast majority of official coordinates worldwide. A few jurisdictions (for example, the Netherlands—EPSG 7415/28992) have adopted oblique stereographic projections as official coordinate systems. Similarly, New Brunswick and Prince Edward Island use separate oblique stereographics (NB, EPSG 2036/2953; PEI, 2292). These, like the transverse Mercator and the Lambert conformal conic, remain conformal with the same error properties.

A few nonconformal projections have been adopted as official coordinate systems. For example, one of the jurisdictions of the United States (Guam) uses an azimuthal equidistant projection (Doyle, 2004) as a part of the U.S. state plane system. Some jurisdictions (Hong Kong, EPSG 3366; Qatar, 2099; Palestine/Israel, 28193; Berlin, 3068) use a Cassini-Soldner projection—a nonconformal transverse Mercator. In most cases, these have been superseded by more standard projections. In the thousands of known systems, the few exceptions can be listed quickly.

These official projections were instituted to reduce the effort for surveyors to convert field measurements onto topographic maps. By choosing conformal projections (Stem, 1989, p. 17), the angles measured in the field need no conversion (though in most cases adjustments for grid north). Distances can be corrected using a single scale factor that varies smoothly in a known manner. The conversion from projection surface to ellipsoid is usually packaged with the local elevation to make a combined scale factor to relate ground measurements to the mapping framework.

Official projections typically use secant geometry, where the reference surface (cylinder or cone) is deliberately smaller than that of the earth. For example, the UTM sets the scale factor at 0.9996 in the central meridian (Figure 3.2). Therefore, distances are –0.04% (short of true) in the neighborhood of the

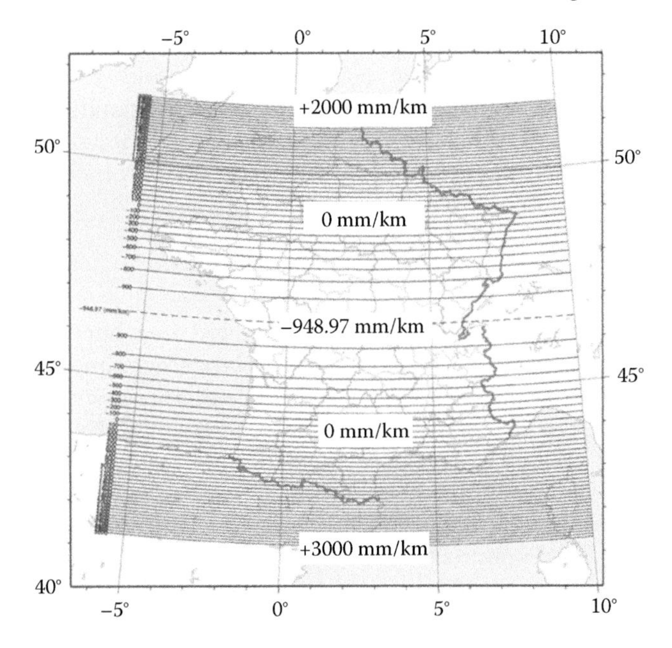

FIGURE 3.2
Scale factor of the RGF Lambert 93 projection. (From IGN-SGN/SICA, Réseau géodésique français, concepts et applications, Institut Géographique National—Service de Géodésie et Nivellement/Service IGN Conseil et Applications, St. Mandé, France, 2009.)

central meridian. For simplicity, we will present the error factor (1-scale factor). For UTM, the error factor starts at –0.0004 on the central meridian of the zone in question. This error reduces outward across the UTM zone (symmetrically) until it reaches 0.0000 at a distance of 180 km approximately. Beyond this distance, the scale factor exceeds 1 so the error is positive and distances are exaggerated. If applied according to the specification (in 6° bands), this error in distance is kept within bounds. However, the balance is maintained mostly at low latitudes. At 50N, the 3° width of the zone is 215.1 km wide, so much more of the zone is underestimated than overestimated.

For more precise work, official coordinate systems typically apply scale factors closer to 1.000, and therefore are limited to smaller regions. The Gauss–Kruger coordinates for Germany (EPSG 31466-9) and a number of other central and eastern European countries adopt 3° wide zones. The scale factor of 1.0000 applies on the central meridian so all distances are overestimated proportionately away from it. Canada adopted MTM, based on 3° zones and a scale factor of 0.9999; hence, there is an error factor of –0.0001 on the central meridian (EPSG 26891-7). The design criteria for the state plane coordinates of the United States specify this same maximum absolute error factor of 0.0001 for each of 127 zones (Stem, 1989, p. 3). For some small states, such as Delaware (EPSG 2776), the scale factor on the central meridian is set at 0.999995. At this level, projection error is not significant for GIS applications.

Some zone systems are designed to cover larger spans than UTM. For example, the TM projection for the UK National Grid (EPSG 27700/7405) is used for the whole of Great Britain and adjacent islands, which extend more than 8° of longitude (Ordinance Survey, 2010). Though at the latitude of northern Scotland, the distance from the central meridian is smaller, and thus the error effect may stay within bounds.

3.2.3 GIS Practices That Increase Error

Zone systems are designed to apply over a limited area, but GIS applications tend to span these zones to encompass bigger landscapes and whole political jurisdictions. GIS databases are intended to be enterprise-wide, not just project based. So, for example, the Wisconsin Department of Natural Resources (DNR) did not operate in the three zones of Lambert conformal conic projections that covered the state in three bands from north to central to south. Instead, they designed their own transverse Mercator with a central meridian of 90W (a zone boundary of UTM). This became a de facto statewide projection system for the state, with a local origin that keeps coordinates in reasonable bounds (now codified as EPSG 3070/3071, called Wisconsin TM). DNR was lucky in that Wisconsin is mostly confined between 87W and 93W, therefore complying with the 6° width of UTM zones. (Only Washington Island off the Door Peninsula is located outside the band, and the overestimate of distances on this island may not adversely influence decision making.) A similar history applies to Idaho, another state more or less confined to 6° east–west and having three zones in the state plane system. In this case, the single zone of TM (using UTM-style parameters) is officially adopted as a state standard (Idaho Technology Authority, 2005).

Similarly, the state of Michigan sought a solution to having three zones of Lambert conformal conic for their oddly shaped jurisdiction. They devised an oblique transverse Mercator that fits the UTM design criteria. With judicious choice of azimuth, they have a single coordinate system for statewide purposes that balances the error factor between +0.0004 and –0.0004 (EPSG 3079). At least 10 jurisdictions in the United States dispensed with multiple zones in the recent overhaul related to the North American datum changes in the 1980s and 1990s (WAGIC, 2010, p. 2; Doyle, 2004).

Other GIS designers are not so lucky. Their jurisdictions are larger, and the drive to have a unitary coordinate system is too strong. They have sometimes taken the strategy of extending coordinate systems beyond their design specifications. For example, many state of Washington agencies use the south zone of their Lambert conformal conic to extend over the whole north zone for statewide use (Washington Information Services Board, 2011, p. 3). On the north border of the state, the error factor in area (not the error in distances) exceeds +0.0017.

Similarly, France has established official coordinates called the grille Lambert, based on the same Lambert conformal conic. The old system had four zones (in east–west bands stacked north–south as conical coordinate systems operate). It was accepted practice to use zone II (centered on Paris, EPSG 27572/7411) extended to the whole of France. As a consequence, the error factor reached +0.005 at 41N (in Corsica). Since 2010, the replacement national system for France is the RGF Lambert 93 projection system (EPSG 2154). This system has error factors of –0.001 in the center of France, +0.002 at Dunkerque in the north, and +0.003 in Corsica (Figure 3.2).

For specific usages of geospatial information that require minimizing the error factor, the addition of local systems was decided. One proposal suggested 39 zones for France to bring the maximum scale error down to 1 cm/km (0.00001). However, the implementation of so many zones was deemed impractical. Finally, 9 zones were selected (EPSG 3942-50; named by their central parallel from 42N to 50 N), in order to bring the maximum local scale error to 0.00008 (Figure 3.3).

The use of extended zones is not limited to Lambert conformal conic projections. Similar issues occur in states (United States) that use the transverse Mercator. There is a temptation to adopt the Illinois west zone to cover the whole state. Proposals for a single optimized projection have been circulated in many states, but the practice is often unregulated. In the conversion to NAD83, a few states (including the relatively large Montana) dropped multiple zones to officially adopt a single projection (Doyle, 2004). This reduces edge issues within the state, but once a study area extends to multiple states, the issues return. In addition, three states now exceed the scale factor limits originally designed for the state plane system (Stem, 1989, p. 16). These

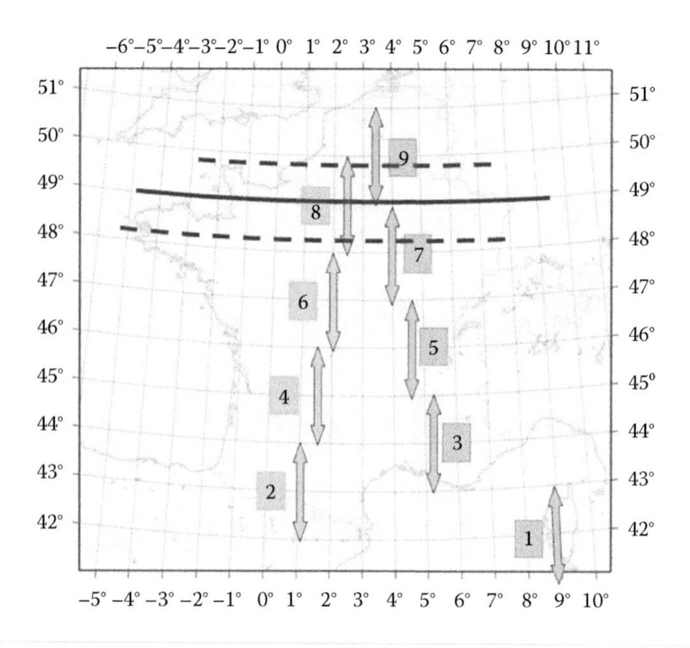

FIGURE 3.3
Layout of the "Lambert 9 zones" conformal conic projections for France used to minimize scale error. (From http://geodesie.ign.fr/index.php?page=documentation.)

concerns are equally relevant in Europe across national boundaries, or in countries like Australia that adopt geometrically simple UTM zones that cut across political jurisdictions.

3.3 Consequences of Projection Error on Analytical Measures

3.3.1 Estimates of Error in Perimeter and Area

Most treatment of projection error stops with the scale factor, the deviation of distances at a point. This factor can be directly applied to measurements of perimeter for polygons or lengths of polylines. Girres (2012, p. 119) performed a test on the Lambert 93 projection for one department in the southwest of France (Pyrénées-Atlantiques) (Figure 3.4). The projection error for the primary road network generates an overestimation of about 0.05% of the total length.

For conformal projections, since they maintain isotropic scale factors, the area error factor e_A can be obtained from the linear error e_L according to Equation 3.1 (Girres, 2012, p. 112; CERTU, 2010, p. 4):

$$e_A = 2 * e_L + {e_L}^2 \tag{3.1}$$

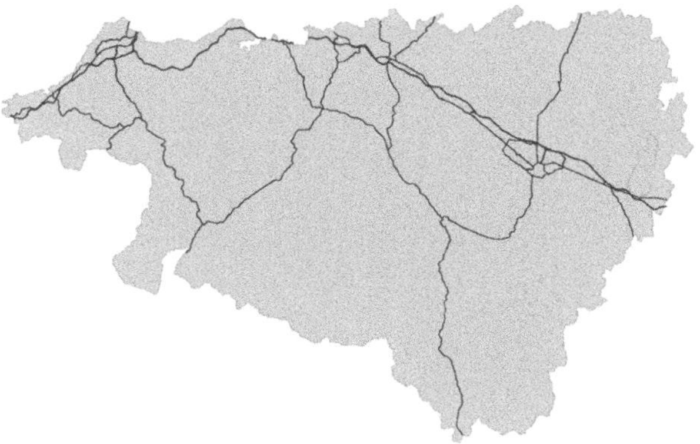

FIGURE 3.4
Primary road network of Pyrénées-Atlantiques. Data measured: IGN-BDTOPO. (From Girres, J.-F., Modèle d'estimation de l'imprécision des mesures géométriques de données géographiques: Application aux mesures de longueur et de surface, PhD, Université Paris–Est, France, 2012, p. 119.)

Therefore, it is approximately double the linear scale error. For the total area of Pyrénées-Atlantiques, the overestimation on the Lambert 93 projection was 0.12%. This figure is consistent with the estimates on the north edge for the state of Washington case. As shown in Figure 3.5, the computation of the ellipsoidal area for the department of Pyrénées-Atlantiques requires oversampling of the polygon (using triangles), since the scale error varies nonlinearly on the entire zone. The area of each triangle was calculated using ellipsoidal trigonometry.

Overall, these figures may seem small, but since they are systematic, they should not occur at all. Figure 3.6 shows a comparison of area calculated for

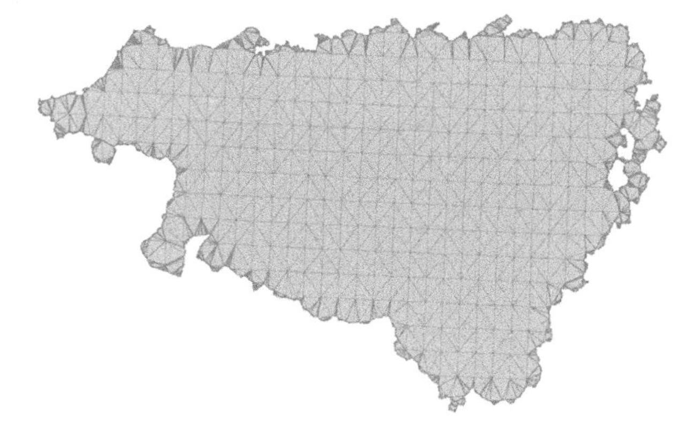

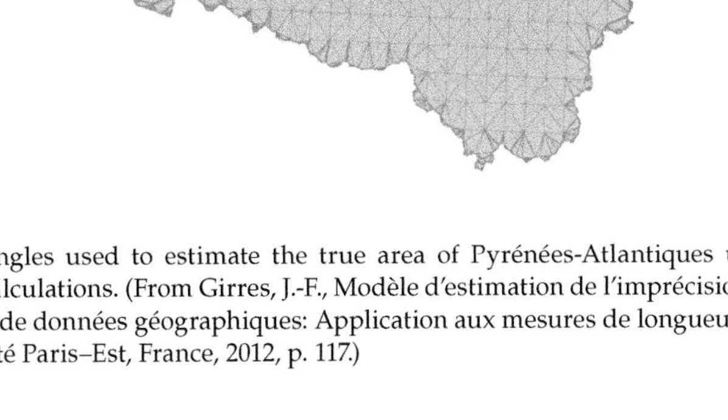

FIGURE 3.5
Spherical triangles used to estimate the true area of Pyrénées-Atlantiques using geodetic (ellipsoidal) calculations. (From Girres, J.-F., Modèle d'estimation de l'imprécision des mesures géométriques de données géographiques: Application aux mesures de longueur et de surface, PhD, Université Paris–Est, France, 2012, p. 117.)

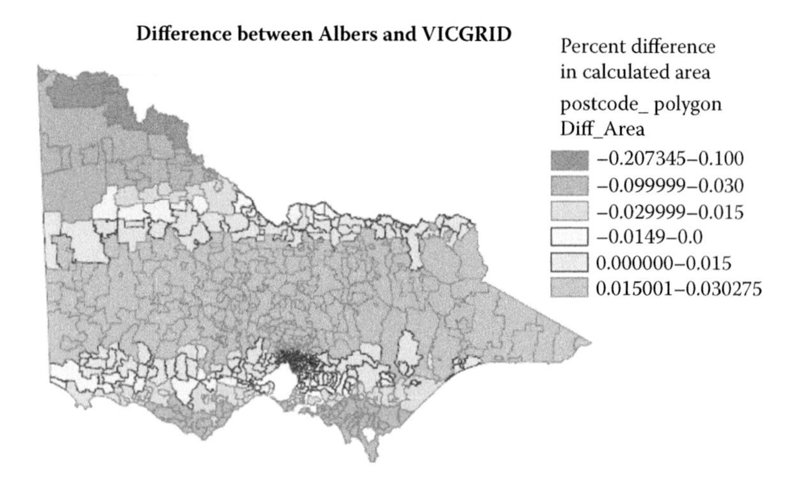

FIGURE 3.6 (See color insert.)
Difference between Albers and VICGRID in percent of area.

the postcode polygons of Victoria, Australia. The official coordinate system VICGRID (a Lambert conformal conic, EPSG 3111) is compared to an Albers projection with the same standard parallels and central meridian. The difference is nearest to zero on the two standard parallels (35S and 38S). Between the parallels, the conformal projection overestimates, and beyond the parallels, the conformal projection underestimates. Northern Victoria, the area farthest from the parallels, has an error factor of −0.002—a result consistent with the results reported in France and Washington state.

A few studies, oriented toward global tessellations, have compared properties of distortion for a number of projections (e.g., Kimerling et al., 1995). The results apply more to the linear scale error than area measurements.

3.3.2 Policy Results

The scale errors reported above can add up to differences in public policy. Harmel (2009, p. 29) reports that France was able to save 17 million euro annually (in the subsidies of the Common Agricultural Policy) after the areas of farm properties were recalculated on the new official RGF Lambert 93 projection, by comparison with the old reference system (zone II extended, shown in green). As shown in Figure 3.7, the error of Lambert 93 projection is systematically inferior to Lambert II (extended) in any latitude and reduces the error factors overall.

Due to repositioning of the standard parallels, the scale factor difference between the two projection systems increases southward. Through these differences, the overestimates by the prior projection system in southern France were significant and cumulative. The area estimates of farm properties

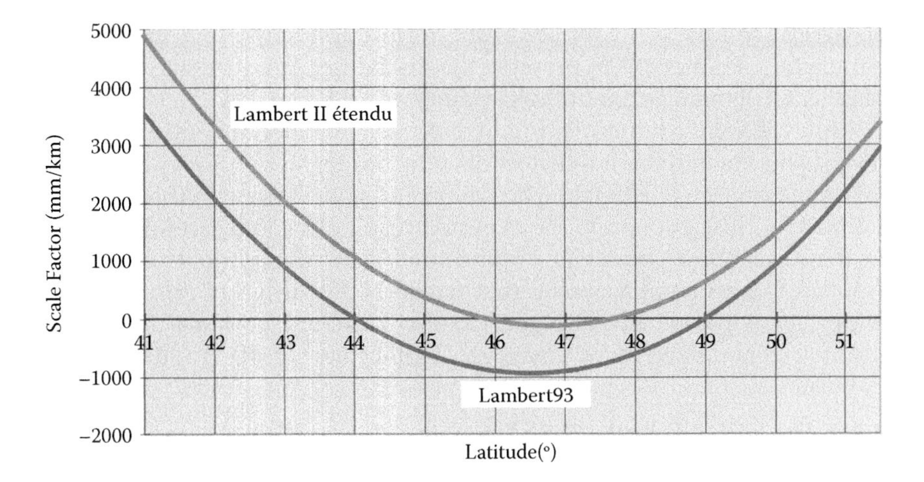

FIGURE 3.7
Scale factor differences between RGF Lambert 93 (bottom line) and NTF Lambert II (top line) projections according to the latitude. (From Harmel, 2009, p. 30.)

revised on the new official projection system (while still subject to some error) were substantially lower. Other situations in other countries may be at least as large, though undocumented as yet. Any subsidy scheme based on farm size would be questionable if not calculated correctly. Similarly, property evaluations based on land area should be reviewed. The oil and gas industry recognizes that area estimates for maritime leases should be done correctly on a geodetic framework (OGP, 2014), perhaps one of the few industries to show some signs of awareness (perhaps due to the money involved).

3.4 Mitigating Error from Projections

As described above, projection errors are systematic, not the result of some random and unknown influence. Thus, they can be predicted with nearly perfect accuracy—in principle. We argue that from an ethical perspective, a professional has the duty to remove any errors that are known. This section will consider solutions to manage this process.

3.4.1 When in Doubt: Disclaimers

A simple solution to the problem of scale error is simply to issue a disclaimer. GIS software could simply state that all calculations were done on a plane projection system that may bear some relationship to the earth, but that numerical results will vary according to certain known (but unexpressed) parameters. Hunter (1999) presents the disclaimer as a first step in the long road to a full management of uncertainty and data quality.

While the legal community might approve, such disclaimers beg the question, if you know there is a distortion, why not correct it? It may make sense to use the projected estimate when the data are so inaccurate that the planar estimate is close enough to the expected result of a more careful treatment. In the cases above, errors in distance and area can rise to 1 part in 1000. Applied to positional accuracy, that would be like a 40 km error on worldwide data, while we would expect close to 1 km (or 1 part in 40,075 for even the most generalized world coverage, Digital Chart of the World).

3.4.2 Use of Local-Scale Error Estimate

The calculations to project a point from geographic (angular) coordinates to the projection can also be solved to provide the scale error at each point (Snyder, 1987; Stem, 1989, p. 18; IGN-SGN, 2010). These values vary systematically and symmetrically from the central meridian of a cylindric projection or the standard parallels of a conic projection. For objects like parcels or urban street segments, the scale error will not change substantially across the object. Therefore, a measure such as perimeter could simply be scaled by the average scale error (or simply a representative value). If the scale error is considered to be an attribute of the feature, then the user can see how the values computed on the projection are scaled in a quite transparent way.

Some objects are large enough that they will have substantial variation internally. The scalar would need to be a weighted average. The exact requirements for this may not be clear in the case of areas when the internal portions of a large polygon may be at a different scale error from the perimeter. Further geometric consideration would have to be undertaken.

3.4.3 Changing Projections

Another solution would be to change from conformal projections. An equal-area (equivalent) projection will provide a solution for area calculations with no rescaling. For instance, the INSPIRE directive recommends the use of the ETRS89 Lambert azimuthal equal-area coordinate reference system for statistical analysis and display reporting in the European Union, where true area representation is required (INSPIRE, 2010) (Figure 3.8).

In the United States, Albers equal-area conic is very commonly used by the U.S. Census Bureau and others as the base for thematic maps. If this projection is used for all area calculations, there will be no distortion. However, this Albers projection is rarely used as the coordinate framework for geodetic control. Snyder (1988) devised other forms of equivalent projections to reduce linear distortion for noncircular areas. Whatever equivalent projection is used, there is no distortion of area measures.

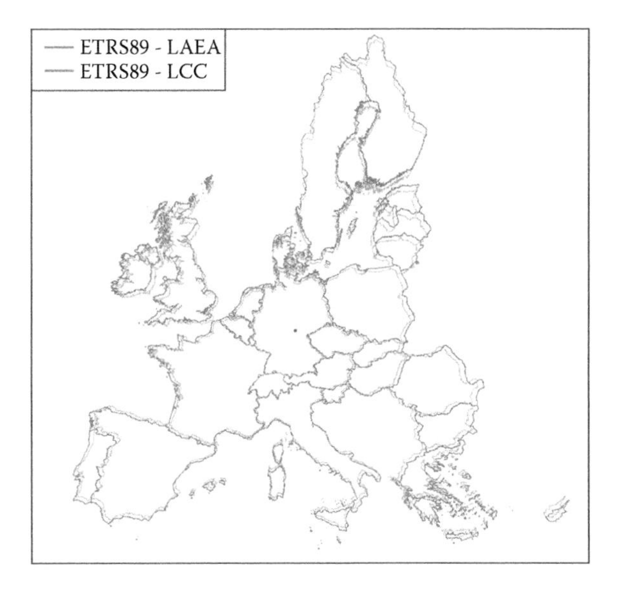

FIGURE 3.8 (See color insert.)
A representation of European Union countries using Lambert azimuthal equal-area (in blue) and Lambert conformal conic (in green) projections (inspired from Dana, 1997). (From http://epp.eurostat.ec.europa.eu/.)

In addition, equivalent projections do not preserve distances (or angles). Therefore, the perimeter of a polygon or any other length measure would still need to be rescaled using the scale error estimates given above. Note that these are more complicated for nonconformal projections in that they are nonisotropic. The distance would have to be recalculated with weighting for each segment, depending on its orientation.

Use of nonstandard projections will require the typical vigilance in integrating sources, but that is a part of normal practice for different data and projection parameters. Additionally, the metadata for a database may have to specify which columns of attributes (such as perimeter or area) were calculated with which projection. Right now we have the naïve assumption that all geometric properties relate to the current projection or to the map view.

3.4.4 Radical Alternative

An alternative must be considered. If all GIS measurements are maintained on the ellipsoid and all analytical results calculated with ellipsoidal trigonometry, there would be no projection error (Lukatela, 2000). The formulae for distances on the ellipsoid are well known and covered in basic geodesy courses (Legendre, 1806; Vincenty, 1975; Deakin and Hunter, 2007). But even these known solutions are often ignored by GIS software designers. The most authoritative recent work on the subject provides nanometer precision (Karney, 2011, 2013), particularly for the inverse geodetic problem—determining distance given two points, but also extending to areas and intersections.

For the case of distance calculations, some widely available software (including Quantum GIS Development Team, 2013) can compute ellipsoidal measures, based on Vincenty's formula (Vincenty, 1975). Unfortunately, on-the-fly computation of geometric length or area on vector objects is still realized using map projections, without considering the scale error. Other software packages may provide specific tools for the geodetic case, but then apply the projection for the more routine geometric processing, including the attributes of perimeter and area attached to each object.

With modern computational capacity, the increased burden of ellipsoidal equations may not be as significant as they were on earlier computing platforms. Our preliminary estimate is that a calculation that takes five floating point operations (probably in double precision) and one square root will be replaced by maybe eight multiplications and four trigonometric function calls.

Areas are not so easy, as Lukatela (2000) has carefully documented. The planar calculation usually divides the polygon into trapezoids (extending to the axis or some reference line). Dividing a polygon on the ellipsoid into pseudotrapezoids based on latitude and longitude leads to objects with complex curves to estimate (e.g., see Balcerzak and Pedzich, 2007; Danielsen, 1989), not the simple subtraction done on the plane. Kimerling (1984) estimated the areas of such trapezoids using approximations linked back to a UTM projection. In geodetic calculation, such approximations are not required. Complex algorithms lead to high-precision results (Danielsen, 1989; Karney, 2013). Lukatela is able to use simpler spherical calculations because the Hipparchus data structure can estimate the local radius of the ellipsoid to provide an adjustment (following the calculation strategy recommended by Legendre, 1806). In the usual representation of angles, this would require trigonometric functions. The Hipparchus system promoted by Lukatela (1987) operates on direction cosines, so these calculations can be greatly simplified by decomposition, either into the geodetic triangles shown in Figure 3.6 or an adaptive Voronoi structure such as that of Hipparchus (Lukatela, 2000).

3.5 Conclusions

This chapter ends on a question of ethics. While some recent work divides ethics into internalist and externalist approaches (Crampton, 1995), it is also germane to consider the potential harm versus the effort required. This utilitarian formulation applies to circumstances in which the person must choose how much effort to expend considering the potential risks.

Most of GIS emphasizes a high degree of spatial precision, sometimes quite unwarranted by the actual accuracy of the data. Areas are reported to six or eight decimal digits of square meters, when the topographic data input may be good to 10 m. Square micrometers or finer results are clearly unjustified. Yet, these highly detailed results may only be accurate to 1 part in 1000 in certain cases of map projections reported above. This level of accuracy means that the numerical results should be rounded to three or four digits, by any routine measurement guideline. Does any software do this? The best way to resolve this problem is to produce calculations based on a more valid model.

In order to reduce the harm done by map projections, GIS professionals have a duty to report measures like perimeter and area after correction for known systematic distortions. In the same way, other well-known sources of error in length and area measures, such as the systematic underestimation involved by the nonconsideration of the relief, should also be estimated and reported to the final user (Girres, 2011, 2012, 2013). Some will argue that these are small effects, but even so, they should not persist.

Acknowledgments

The doctoral research of J.-F. Girres was supported by IGN, under the direction of Anne Ruas and Université Paris–Est. Chrisman acknowledges support from RMIT University for travel funds.

References

Balcerzak, J., and Pedzich, P. 2007. The methods of calculation of ellipsoidal polygon areas based on some map projection properties. Presented at Proceedings FICC 2007, paper 2.2.3. http://icaci.org/files/documents/ICC_proceedings/ICC2007/documents/doc/THEME%202/oral%202/2.2.3%20THE%20METHODS%20OF%20CALCULATION%20OF%20ELLIPSOIDAL%20P.doc.

CERTU. 2010. Distances et surfaces, comment prendre en compte l'altération linéaire? In *Géoréférencement et RGF93, Théorie et concepts—Fiche T9*, pp. 1–6. Lyon, France. http://www.certu.fr/fr/_Information_géographique-n32/Géoréférencement_et_RGF93-n795/IMG/pdf/RGF93_theorie_et_concept_T9.pdf.

Crampton, J. 1995. The ethics of GIS. *Cartography and Geographic Information Systems*, 22(1), 84–89.

Dana, P.H. 1997. The geographer's craft project. Department of Geography, University of Colorado, Boulder. http://www.colorado.edu/geography/gcraft/notes/mapproj/gif/threepro.gif.

Danielsen, J.S. 1989. The area under the geodesic. *Survey Review*, 30(232), 61–66.

Deakin, R.E., and Hunter, M.N. 2007. Geodesics on an ellipsoid: Pittman's method. In *Proceedings of the Spatial Sciences Institute Biennial International Conference (SSC2007)*, Hobart, Tasmania, Australia, May 14–18, 2007, pp. 223–242.

Doyle, D.R. 2004. Computing state plane coordinates. *Professional Surveyor*, 24(1). http://archives.profsurv.com/magazine/article.aspx?i=1180.

Girres, J.-F. 2011. A model to estimate length measurements uncertainty in vector databases. In *Proceedings of the 7th International Symposium on Spatial Data Quality*, Coimbra, Portugal, pp. 83–88.

Girres, J.-F. 2012. Modèle d'estimation de l'imprécision des mesures géométriques de données géographiques: Application aux mesures de longueur et de surface. PhD, Université Paris–Est, France.

Girres, J.F. 2013. Estimation de l'imprécision des mesures géométriques de longueur et de surface: Proposition d'un modèle destiné aux utilisateurs de données géographiques. *Comité Française de la Cartographie*, 215, 45–50. http://www.lecfc.fr/new/articles/215-article-8.pdf.

Griffith, D.A. 1989. Distance calculations and errors in geographic databases. In M.F. Goodchild and S. Gopal (eds.), *The Accuracy of Spatial Databases*, pp. 81–90. Taylor & Francis, London.

Harmel, A. 2009. Le nouveau système réglementaire Lambert 93. *Géomatique Expert*, 68, 26–30. http://www.geomag.fr/sites/default/files/68_91.pdf.

Hunter, G.J. 1999. Managing uncertainty in GIS. In D. Maguire, M.F. Goodchild, and D. Rhind (eds.), *Geographical Information Systems*, Vol. 2, pp. 633–641. Longmans, London. http://www.geos.edac.uk/gisteac/gis_book_abridged/files/ch45.pdf.

Idaho Technology Authority. 2005. S4210: Single zone coordinate system for GIS data. In Enterprise standards: S4200. http://ita.idaho.gov/psg/s4210.pdf.

Institut Géographique National—Service de Géodésie et Nivellement (IGN-SGN). 2010. Circé france version 4. IGN-SGN, St. Mandé, France. http://geodesie.ign.fr/contenu/fichiers/Plaquette_CirceFrance_v4.pdf.

Institut Géographique National—Service de Géodésie et Nivellement/Service IGN Conseil et Applications (IGN-SGN/SICA). 2009. Réseau géodésique français, concepts et applications. IGN-SGN/SICA, St. Mandé, France.

INSPIRE. 2010. Infrastructure for spatial information in the European community. In *INSPIRE Specification on Coordinate Reference Systems: Guidelines*. INSPIRE Thematic Working Group on Coordinate Reference Systems and Geographical Grid Systems. http://inspire.jrc.ec.europa.eu/documents/Data_Specifications/INSPIRE_Specification_CRS_v3.1.pdf.

Karney, C.F.F. 2011. Geodesics on an ellipsoid of revolution. arXiv: 1102.1215. http://arxiv.org/pdf/1102.1215.pdf.

Karney, C.F.F. 2013. Algorithms for geodesics. *Journal of Geodesy*, 87(1), 43–44. http://link.springer.com/article/10.1007%2Fs00190-012-0578-z.

Kimerling, J.A. 1984. Area computation from geodetic coordinates on the spheroid. *Surveying and Mapping*, 44(4), 343–351.

Kimerling, J.A., Overton, S.W., and White, D. 1995. Statistical comparison of map projection distortions within irregular areas. *Cartography and Geographic Information Systems*, 22(3), 205–221.

Legendre, A.M. 1806. Analyse des triangles tracés sur la surface d'un sphéroïde. In *Mémoires de la classe des sciences mathématiques et physiques de l'Institut Nationale de France*, premier semestre, pp. 130–161, http://books.google.com/books?id=-d0EAAAAQAAJ&pg=PA130-IA4.

Leung, Y., Ma, J.H., and Goodchild, M.F. 2004. A general framework for error analysis in measurement-based GIS. Part 4: Error analysis in length and area measurements. *Journal of Geographical Systems*, 6, 403–428.

Lukatela, H. 1987. Hipparchus geopositioning model: An overview. In *Proceedings AUTO-CARTO 8*, pp. 87–96. http://mapcontext.com/autocarto/proceedings/auto-carto-8/pdf/hipparchus-geopositioning-model-an-overview.pdf.

Lukatela, H. 2000. Ellipsoidal area computations of large terrestrial objects. In M.F. Goodchild and J.A. Kimerling (eds.), *Discrete Global Grids*. Santa Barbara, CA: National Center for Geographic Information and Analysis (NCGIA). http://www.ncgia.ucsb.edu/globalgrids-book/eac/.

International Organization of Oil and Gas Producers (OGP). 2009. Geomatics guidance note 5: Coordinate reference system definition. OGP report no. 373-05. http://www.ogp.org.uk/pubs/373-05.pdf.

International Organization of Oil and Gas Producers (OGP). 2014. Geomatics guidance note 3: Contract area description. OGP report no. 373-03. http://www.iogp.org/pubs/373-03.pdf.

Ordnance Survey. 2010. A guide to coordinate systems in Great Britain, OS, Southampton, UK. http://www.ordnancesurvey.co.uk/oswebsite/gps/docs/A_Guide_to_Coordinate_Systems_in_Great_Britain.pdf.

Quantum GIS Development Team. 2013. Quantum GIS geographic information system. Open Source Geospatial Foundation Project. http://qgis.osgeo.org.

Shi, W., and Liu, W. 2000. A stochastic process-based model for the positional error of line segments in GIS. *International Journal of Geographical Information Science*, 14(1), 51–66.

Sillard, P. 2000. *Les projections et référentiels cartographiques*. Ecole Nationale des Sciences Géographiques, Marne-la-Vallée, France. http://fad.ensg.eu/moodle/file.php/11/Geodesie/projections.pdf.

Snyder, J.P. 1987. Map projections: A working manual. Professional Paper 1395. U.S. Geological Survey, Reston, VA. 3rd rev. printing 1994. http://pubs.usgs.gov/pp/1395/report.pdf.

Snyder, J.P. 1988. New equal-area projections for noncircular regions. *American Cartographer*, 15(4), 341–356.

Stem, J. 1989. State plane coordinate system of 1983. NOAA Manual NOS NGS 5. National Geodetic Survey, Rockville, MD. http://www.ngs.noaa.gov/PUBS_LIB/ManualNOSNGS5.pdf.

Vincenty, T. 1975. Direct and inverse solutions of geodesics on the ellipsoid with application of nested equations. *Survey Review*, 22, 88–93. http://www.ngs.noaa.gov/PUBS_LIB/inverse.pdf.

Washington Geographic Information Council (WAGIC). 2010. Adopting a single state plane coordinate system: Call to action. Draft document (proposal not implemented). http://geography.wa.gov/wagic/zonetostate.pdf (page no longer online).

Washington Information Services Board. 2011. Geographic information systems (GIS) geodetic control standards. Standard 601-S2. http://ofm.wa.gov/ocio/policies/documents/161.10.pdf.

4

Function Workflow Design for a Geographic Information System: A Data Quality Perspective

J.-H. Hong and M.-L. Huang

Department of Geomatics, National Cheng Kung University, Tainan City, Taiwan

CONTENTS

ABSTRACT Although data quality has been well accepted as an essential component of geospatial data, it has not received its deserved attention in geographic information system (GIS)–based applications over the years. Users are forced to deal with data of unknown or ambiguous quality, and thus unpredictable risk in decision making is inevitable. Quality information must be fully assimilated into the design of GIS functions and further extended to the visualization of processing results. This requirement is particularly necessary for users who do not possess the required knowledge to correctly interpret the outcomes of GIS functions. We propose a quality-aware workflow design approach driven by standardized quality information and then use the data select-by-region function as an example to demonstrate how quality information changes the traditional design principles of GIS functions. The proposed workflow not only improves the interoperability of heterogeneous geospatial data but also significantly increases the intelligence of GIS-based functions.

4.1 Introduction

Although data quality has long been accepted as an essential component of geospatial data, as indicated by the history of spatial data quality research in the work of Devillers et al. (2010), proper use of the acquired geospatial data remains a significant challenge for current geographic information system (GIS) users. The ability of GISs to integrate data of various qualities can lead to errors in the final results (Lanter and Veregin, 1992). Traditional GISs rely heavily on experts to understand the data they use and ensure that every decision is made correctly. As the sharing of geospatial data has become easier with the emergence of geoportals (Maguire and Longley, 2005), the implicit or explicit discrepancy and heterogeneity of data quality between data sets acquired from various georesources have become major obstacles to data interoperability. Given the lack of a comprehensive framework for modelling, distributing, analyzing, and visualizing the quality of heterogeneous geospatial data, GIS users are forced to deal with data of ambiguous or even unknown quality. Thus, an unpredictable level of risk that users may never even notice is inevitably hidden in the final decisions. Given that GIS functions are often naively used to process, analyze, and derive new information, we argue that the awareness of data quality must be considered during the design of GIS functions. Otherwise, users will be working in an extremely risky application environment and cannot take full advantage of the resource sharing brought by spatial data infrastructure (SDI).

Geospatial data quality serves as the basis for determining its fitness for use in a particular application (Devillers et al., 2007b). For example, topographic mapmaking follows rigorous mapping specifications to ensure the quality of the final product. Because different scales of topographic maps represent different levels of abstraction and quality, users are trained to select the right scale of such maps according to their application needs. For professional use, the mapping specifications can serve as a reference to obtain more details about how topographic map data are selected, observed, processed, and mapped. The fitness-for-use decision becomes difficult if quality information is not available. Thus, the development of domain-specific or universal standards for quality information has been a major requirement in the cartography and GIS domains. Recently, ISO/TC211 published the international standards ISO 19113 (ISO, 2002), 19114 (ISO, 2003a), and 19138 (ISO, 2006) to address the issues of principles, evaluation procedure, and data quality measurement, respectively. This standardized framework provides a consensus agreement on the measurement of spatial data quality for various types of GIS data. By including a standardized framework of quality information in the ISO 19115 metadata standard (ISO, 2003b), data distributors can store, maintain, and provide access to the metadata that describe the data quality, licensing, and pricing properties (Dustdar et al., 2012). To facilitate geospatial data sharing, the standardized metadata serves as a solid foundation for building an SDI.

When dealing with GIS data, users should be cautiously and constantly aware that the contents of data sets are generated according to their original needs. The inability of different data sets to merge due to the quality issue is not surprising. Users are certainly expected to have an accurate understanding of the data they use, but this understanding can be accomplished only when quality information is available and users have knowledge to interpret the data correctly. Goodchild (2004) proposed the concept of measurement-based GIS, in which the measurement details are retained to propagate the error of position whenever necessary. The concept of the quality-aware GIS (Devillers et al., 2005; Yang, 2007; Devillers and Zargar, 2009) is intended to include the consideration of data quality within the decision-making process. Rather than wait for experts to inspect the quality of selected data sets individually, the quality-aware GIS with built-in knowledge automatically prompts with useful information to aid in user decision making. Devillers et al. (2007a) and Yang (2007) transformed the data quality information into symbols to illustrate their differences in the map interface. Zargar and Devillers (2009) modified the measure operation in ArcGIS to demonstrate that the inclusion of quality reporting (position accuracy, completeness, and logical consistency) can improve the quality of decision making. Hong and Liao (2011) proposed the theory of valid extent to graphically illustrate the data completeness status of multiple data sets in the map interface. With the rapid growth and availability of quality information following metadata standards, the role of quality information is expected to be more versatile in future GIS-based applications. This chapter aims to propose a new idea for the design of GIS functions based on standardized metadata, quality-aware algorithms, and visual aids of quality information.

The rest of this chapter is organized as follows. Section 4.2 explores the relationship between quality information of geospatial data and GIS functions. Section 4.3 proposes the encoding strategy for geospatial data and quality information. Section 4.4 presents the general workflow for implementing the quality-aware GIS. In this section, a case using the data select-by-region function is discussed to demonstrate the difference after data quality information is considered. Finally, Section 4.5 concludes with our major findings.

4.2 Required Data Quality for GIS Functions

Each GIS is a toolbox consisting of hundreds of powerful functions that can handle a variety of geospatial application needs. Each GIS function has its own designed purpose and limitation, which serve as the basis for designing its input, output, and algorithms. GIS professionals are trained to select the right functions and right data to solve the problems at hand under the right conditions. However, users typically assume that data sets are perfect for the conditions of the designed functions without considering their discrepancies in quality. In addition to the display of metadata, current GISs support extremely limited functions regarding the interpretation and evaluation of data quality. For example, awareness of the differences of time, accuracy, scale, and criteria of data selection is necessary when overlaying different data sets, but users are presented only with a superimposed result of selected data sets for visual inspection, with no other information to indicate their differences in quality. Without appropriate aids, such a result may be misleading and even cause significant errors in decision making.

By considering data quality, we can increase the intelligence of GIS functions to avoid poor decision making. In this study, we select three types of GIS functions—conversion, measurement, and selection—and examine the distinguishing characteristics of each. The analysis then serves as the basis for determining the type of quality information that must be considered during the function design process. Four types of data quality elements from the ISO 19113 framework were selected for analysis: completeness, logical consistency, positional accuracy, and attribute accuracy. Although temporal accuracy was ignored in this study, each data set has an implicit temporal property represented by the time the data set was collected. When more than one data set is chosen, GIS functions should automatically analyze the temporal difference of the selected data sets because the quality evaluation is valid only at that particular time instance. For time-sensitive GIS functions, to assume naively that all data sets refer to the same time is inappropriate. For example, the distance measurement between two moving features is meaningless when their temporal information is not considered. The same principles can apply to any GIS function that is based on distance measurement, for example, the buffer.

The following sections examine the characteristics of the three selected categories of GIS functions and their corresponding quality elements.

4.2.1 Conversion

A conversion function changes the content of a data set from its original status to the desired status. Although users are interested in the results after conversion (sometimes only the illustrated result), they tend to ignore how the data set quality changes due to the conversion function. Coordinate

TABLE 4.1

Quality Elements to Consider for Conversion Functions

Quality Element	Evaluation
Data completeness	Whether the number of features remains the same
Positional accuracy	Positional accuracy of conversion results
Logical consistency	Feature dimensionality
	Required topological relationship
Lineage	Conversion method

transformation is a typical conversion function that transforms the geographic coordinates from their originally referenced coordinate system to those of another system to adapt to particular application needs. Although the function can be easily executed after specifying the data set and the target coordinate system, a critical issue is that the positional accuracy after coordinate transformation largely depends on the transformation methods used. The positional accuracy may deteriorate significantly if an approximate transformation method is used. Ignoring the influence of transformation methods is highly risky, and using the positional accuracy report of the source data set as the quality basis of the newly transformed coordinates is even worse. Although a user-friendly GIS working environment allows the coordinate transformation to be easily completed with only a few mouse clicks, such functions normally work as a black box by hiding the implementation details from users. Users can input data sets, specify requests, and receive outputs, but they are not allowed to intervene in the transformation process.

After the conversion process, failure to update the data quality status in metadata and to present appropriate explanations to users may cause unpredictable errors. The conversion result (coordinates) may look appealing, but the transformation history is not provided (lineage and positional accuracy changes). Depending on the purpose and types of conversion, the corresponding data quality elements must be evaluated and added to the design principle. Table 4.1 summarizes the data quality considerations for designing conversion functions.

4.2.2 Measurement

Measurement functions provide tools for users to measure the geometric properties of selected features (e.g., distance and area). Simply providing the measurement values does not help users to evaluate how reliable these values are. Because the measurement is based on the location of features, the positional accuracy information serves as the basis for evaluating the quality of measurement results. Some measurement functions allow users to digitize selected points in the map window visually. Under such circumstances, not only the positional accuracy of the features, but also how these

TABLE 4.2

Quality Elements to Consider for Measurement Functions

Quality Element	Evaluation
Positional accuracy	Error propagation of measurement
	Positional accuracy of reference data
Lineage	Conditions for acquiring coordinates

features are presented to users (e.g., zoom level) must be considered in the working environment and included in the lineage information. From a temporal perspective, if users are measuring two features acquired at different times, a warning message should prompt users that these two features are not guaranteed to coexist. The information on the temporal difference can help determine whether the measurement is a meaningful action. Table 4.2 summarizes the necessary considerations of measurement functions.

4.2.3 Selection

Selection functions allow users to retrieve a subset of features that meet the specified constraints of users. Because the queried result depends on the comparison of data sets and given constraints, the quality of the data greatly influences the results. In practice, users appear to ignore the fact that data are not perfect and make decisions simply based on the selected results. Thus, completeness of the data set must be considered for all of the selection functions because it describes the difference between the data sets and reality. For selection functions based on geometric constraints, positional accuracy and topological consistency must be considered. For example, the "touch the boundary of" function is based on the mathematical formalization of the topological relationship between two features. Unless the data are created following rigorous topological constraints, one feature seldom touches the other feature due to reasons of quality. For the "select by thematic constraint" function, the results depend on the thematic accuracy of the attributes. Table 4.3 summarizes the necessary considerations for selection functions.

TABLE 4.3

Quality Elements to Consider for Selection Functions

Quality Element	Evaluation
Data completeness	Omission and commission measure of queried data
	Spatial extent of the quality evaluation result
Positional accuracy	Positional accuracy of queried data
	Positional accuracy of query constraints
Logical consistency	Topological consistency of queried data
Thematic accuracy	Thematic accuracy of attributes used as constraints
Lineage	Functions and selection constraints

TABLE 4.4

GIS Functions and Related Quality Elements

Category	Function	Completeness	Logical Consistency	Positional Accuracy	Thematic Accuracy
Conversion	Feature to raster	M	O	M	M
	Raster to feature	M	M	M	M
	Coordinate transformation	M	O	M	×
Measurement	Area measurement	×	O	M	×
	Distance measurement	×	O	M	×
Selection	Select by location	M	O	M	×
	Select by attribute	M	O	×	M

Note: M, mandatory; O, optional; ×, not necessary.

Table 4.4 summarizes the selected GIS functions and their required data quality elements. M denotes quality elements highly suggested to be included in the design, O denotes elements that can be included under special conditions, and × denotes elements that need not be considered.

Although GIS functions have been used for quite some time, they do not provide useful information to support correct use and informed decision making. Functions included in a similar category are expected to often share similar design principles of data quality. However, as every function has its own unique characteristics, the modified workflow and corresponding data quality elements must be examined individually. After data quality is considered in the function design, the interpretation of similar outcomes may be extremely different.

4.3 Data Quality Encoding Strategy

Unless data providers are willing to supply quality information, end users can seldom gain a comprehensive understanding of the acquired data. Thus, an essential requirement for a quality-aware GIS is the standardized encoding and interpretation of quality information, which requires a link between the distributed geospatial data and the metadata. The open data and service architecture from the Open Geospatial Consortium appears to be a good candidate for distributing these two types of information due to their standardized and Extensible Markup Language (XML)–based nature. The following subsections discuss the design strategies to integrate XML-based geospatial data and quality information.

4.3.1 Data Quality Scope

Every quality description theoretically has its own data scope, which represents the domain of the data from which the quality information is evaluated. This idea implies that quality information is valid only within this specified scope. According to ISO 19113, this scope information must be unambiguously specified for every individual quality evaluation result. In ISO 19115, four major types of scope—data set series, data set, feature, and attribute—are identified according to the target of the evaluation procedure. A hierarchical relationship exists among these four data scopes. A data set is composed of a number of features; thus, the data quality information can be recorded at the data set level if the evaluation procedure for all of the features in the entire data set is consistent and all the features share similar quality content information. This property simplifies the encoding of data quality information and prevents its duplication at the feature level. However, if the data quality of individual features in the data set differs from one to another (e.g., positional accuracy), then the quality information would be recorded at the feature level for unambiguous interpretation.

Following common practices for establishing metadata, most of the quality information is created by referring to individual data sets. However, quality information may also refer to the level of data set series, feature, or attribute under certain conditions. For example, the quantitative measures of data completeness are based on the omission error and commission error after comparing the data set with the universe of discourse, so the data scope refers by default to a single data set. Conversely, the positional accuracy may refer to either a data set or a feature, depending on the positioning technology and surveying procedures used. Recording metadata at the data-set level often implies that the positional accuracy of features complies with the rules from the data-set specification or is quantified using measures such as root mean square error. When recording at the feature level, data distributors intend to emphasize the difference in positional accuracy between individual features. Regardless of the selected scope for recording quality information, this information must be encoded via a standardized framework such that the application program can unambiguously parse and interpret the distributed quality information.

4.3.2 Encoding Strategy of Data and Quality

In this study, the distributed geospatial data and their related quality information are encoded in Geography Markup Language (GML) and XML following the ISO 19136 and 19139 standard schemas, respectively. The open encoding framework allows client applications to parse necessary temporal,

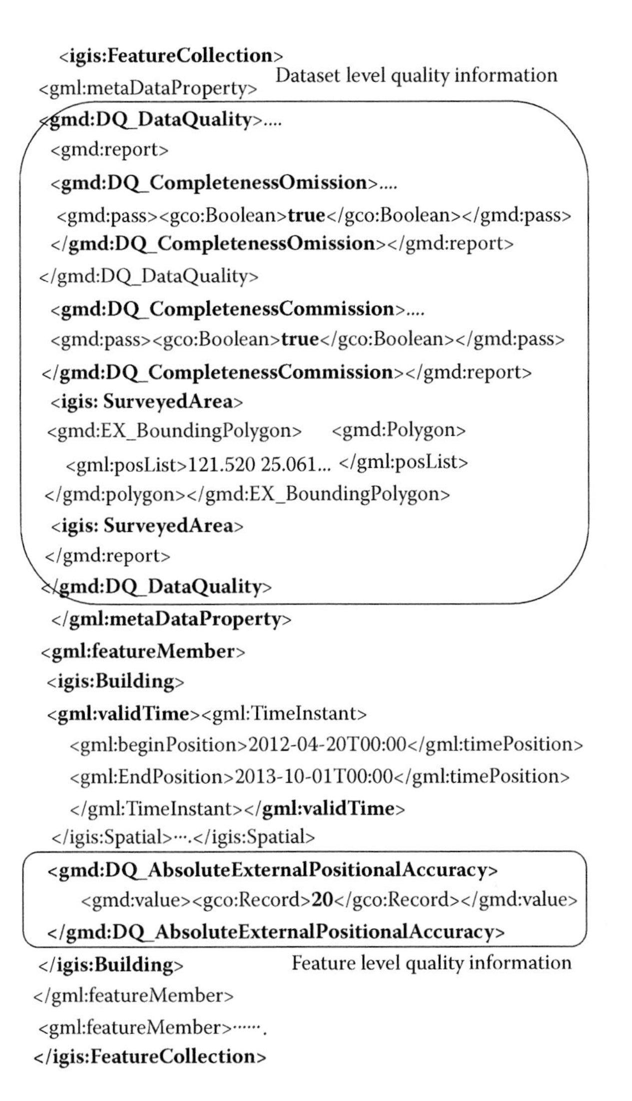

FIGURE 4.1
XML encoding example of the "building" data set.

geometric, attribute, and quality information transparently based on individual features for further processing. Figure 4.1 shows an XML encoding example of the data set "building," which comprises a number of buildings, each of which is encoded as a "featureMember." Because the data completeness evaluation is based on the entire data set, the evaluation result is recorded at the data-set level by the tag "metaDataProperty." In addition to the quantitative measures of metadata elements, the tag of "SurveyedArea" (refer to discussion in Section 4.4.1.2) is an expanded element based on the suggestion of Hong and Liao (2011). Conversely, the positional accuracy information is referred to the feature level and designed as attributes of the building class, following the ISO 19139 schema.

4.4 Intelligence for GIS Function Design

One distinct difference between the traditional GIS and the quality-aware GIS is that the former deals with the data only, whereas the latter helps users evaluate the difference between the results and reality. This section introduces the general workflow for designing quality-aware GIS functions and demonstrates their major differences from the current design principles.

4.4.1 Workflow Rule of Quality-Aware GIS Operations

The design of GIS functions typically involves three components: input, algorithm, and output. The input component demands both the geospatial data and their metadata, as described in Section 4.3. Compared with the current workflow, the improved algorithm additionally includes the evaluation and constraints of data quality according to the purpose of the function. Finally, the output component must be augmented with a new design capable of visualizing the data quality status of the results. The entire process is referred to as being automatic, that is, not requiring user intervention. An ideal scenario is when users can still execute GIS functions in the manner with which they are already familiar; the procedures for acquisition, parsing, interpretation, processing, and visualization are all executed automatically behind the scene.

In the following discussion, we choose the commonly used data select function as an example to explain the design of quality-aware functions. Data select functions vary, depending on the types of constraints (e.g., location, topological, temporal, and attribute). All GIS functions included in this family share a similar purpose of filtering out the data that fail to pass the given constraint and responding to users with only the data that meet given constraints. The selected result only reflects what is being recorded in the data set. Users can safely use the result only after understanding its difference with respect to the reality; that is, quality consideration is necessary.

4.4.1.1 Input

When executing a data select function, users have to first specify the target data sets. The input component is responsible for parsing the spatial, temporal, identification, and quality information of features. Any missing information must be carefully and automatically identified and prompted to users for further reference. For example, the data completeness information serves as the basis for evaluating whether all of the queried features within the spatial constraint (e.g., buffer) are found. If this condition cannot be ensured, consequent functions, such as feature count or attribute statistics, may contain fatal errors. With prompted information, users can deselect data sets that lack data completeness information, or they can assess the risk of using those data sets before taking any further action.

4.4.1.2 Algorithm

Because different types of data quality elements are necessary for different GIS functions, the algorithm design must follow a customized approach. Each data select function should consider the influence of the data completeness status of the queried data set. Hong and Liao (2011) proposed a formal method to geographically describe and illustrate the data completeness status. Every data set is assumed to have a surveyed area, which denotes a spatial extent from which the data are collected. The surveyed area is different from the spatial extent defined via the commonly used minimum bounding box approach, which is often regarded as the approximate spatial representation of the features. In Figure 4.2, the dashed line represents the surveyed area and the solid line represents the minimum bounding box of the features. We can further deduce the major difference that no feature located within the shaded area exists if the information of surveyed area and completeness measure is available.

Based on the above-mentioned surveyed area and completeness measure, the design of the select-by-region function theoretically requires the queried region to be completely within the surveyed area of the data set to ensure that all of the features within the queried region are returned. Otherwise, the returned result may represent only part of the data within the specified region, making statistical reports misleading. Figure 4.3 illustrates three scenarios. The triangle symbols represent features within a selected data set, the polygons of solid lines represent the spatial query, and the polygons of dashed lines represent the surveyed area of the data set. If we assume that no omission and commission errors are found, then the queried results should

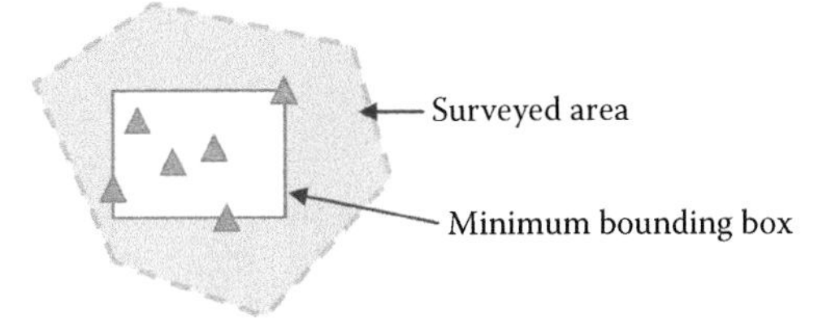

FIGURE 4.2
Concept of the surveyed area and the minimum bounding box of features.

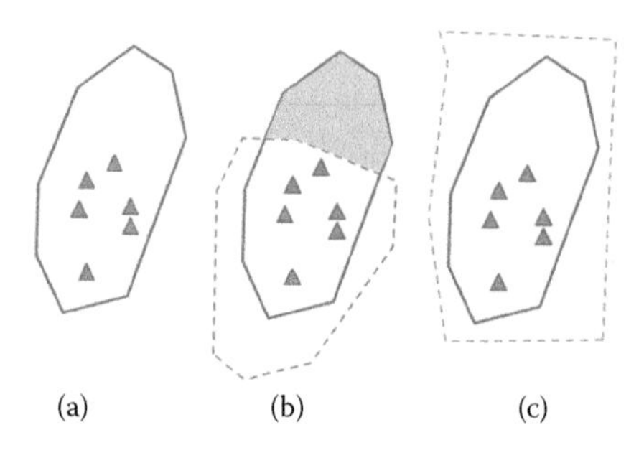

FIGURE 4.3
Select-by-region function. (a) No information about the surveyed area is available. (b) The surveyed area overlaps the queried region. (c) The queried region is within the surveyed area.

be the same in the three cases. However, their interpretations are rather different. In case (a), as no information about the surveyed area is available, we know that seven features are found, but we are not sure if only these seven features are within the queried region. In case (b), we conclude that only seven features exist within the bottom half of the queried region, but the status for the area in gray is unknown. Possibly, either no features or numerous features are located within that region. As no information for the gray area is available, drawing any meaningful conclusion is impossible. Finally, in case (c), we can safely conclude that only seven features exist within the queried region. The availability of surveyed area information does make a difference when executing the select-by-region function. If omission and commission errors are found, then some features may possibly still be missing or wrongly created. Thus, assuming that all the features are available when in fact they are not is inappropriate.

Furthermore, the topological relationships between the queried region and features determine the result of the select-by-region function, so the positional accuracy of data sets and spatial constraints must also be considered. Finally, conducting a spatial query may be meaningless if the validity time of the features and queried region is different, so the temporal difference also has to be considered. All of the aforementioned factors that may influence the outcome of the results must be analyzed from a data quality perspective and unambiguously prompted to users with the appropriate visual aid technique.

4.4.1.3 Output

The output component is responsible for providing useful textual or visual aids to prompt users about the data quality status of the result of GIS function execution. For the select-by-region function, the geometric intersection of the queried region and the surveyed area of the data set must be provided to users for visual inspection. Particularly for regions that are a part of the queried region but outside the surveyed area of the data set, the function represents an area where no information is available. This visual approach subdivides the map interface into a number of regions with different data quality levels so that users are always aware of any possible risks when making decisions.

By incorporating data quality information into the workflow design of GIS functions, we add a new perspective to the design of intelligent GIS functions. The quality-aware result provides a reasonable evaluation of the actual situation. The selection of necessary data quality elements for an individual GIS function depends on its unique purpose and characteristics. For example, the design of the select-by-region function must consider data completeness and positional accuracy, whereas the select-by-attribute function must consider both data completeness and attribute accuracy. Figures 4.4 and 4.5 present the modified workflows of the select-by-region and select-by-attribute functions, respectively. These workflows follow a similar design concept, but each has its own unique algorithm for addressing data quality.

Figure 4.6 summarizes the quality-aware approach discussed in this chapter. By adding the consideration of standardized quality information, the design of the input, algorithm, and output of the GIS functions requires a new perspective.

4.4.2 Use Case

The method was tested through a use case for the commonly used select-by-region function to demonstrate the advantages of adding quality consideration into the design process. The evacuation plan for chlorine gas exposures from a semiconductor company at Tainan Science Park in Tainan City, Taiwan, was selected as the test scenario. The Tainan Science Park is located in an area between the Xinshi, Shanhua, and Anding districts of Tainan City, covering a total of 2578 acres. When the information on the threat zone is available, the most straightforward approach for the evacuation mission is to use the selection-by-region function on the relevant data sets (e.g., buildings, schools, and factories). This function is typically considered a geometric function for which features are added to the result if their locations are within the threat zone. All of the tests were developed using Visual Basic and ESRI ArcGIS 10 software. We further used the Area Locations of Hazardous Atmospheres (ALOHA) software to simulate the spatial extent of the threat zone. ALOHA is an air dispersion model used to predict the movement and dispersion of gases (USEPA and NOAA, 1999). This software calculates the downwind dispersion of a chemical cloud based on the toxicological and physical characteristics of the released chemical, atmospheric conditions, and specific circumstances. Figure 4.7 illustrates the threat zone of chlorine gas exposures calculated using the ALOHA software.

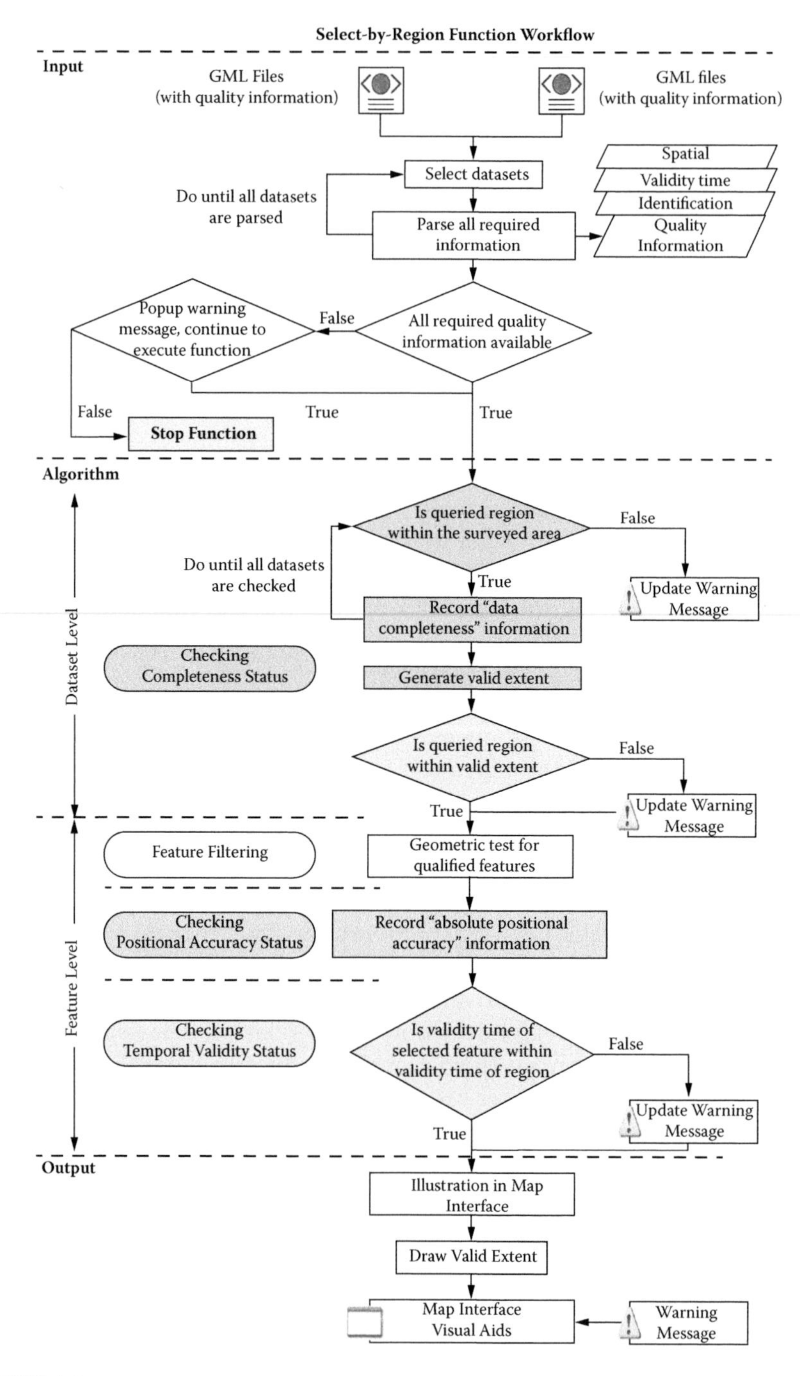

FIGURE 4.4
Workflow of the select-by-region function.

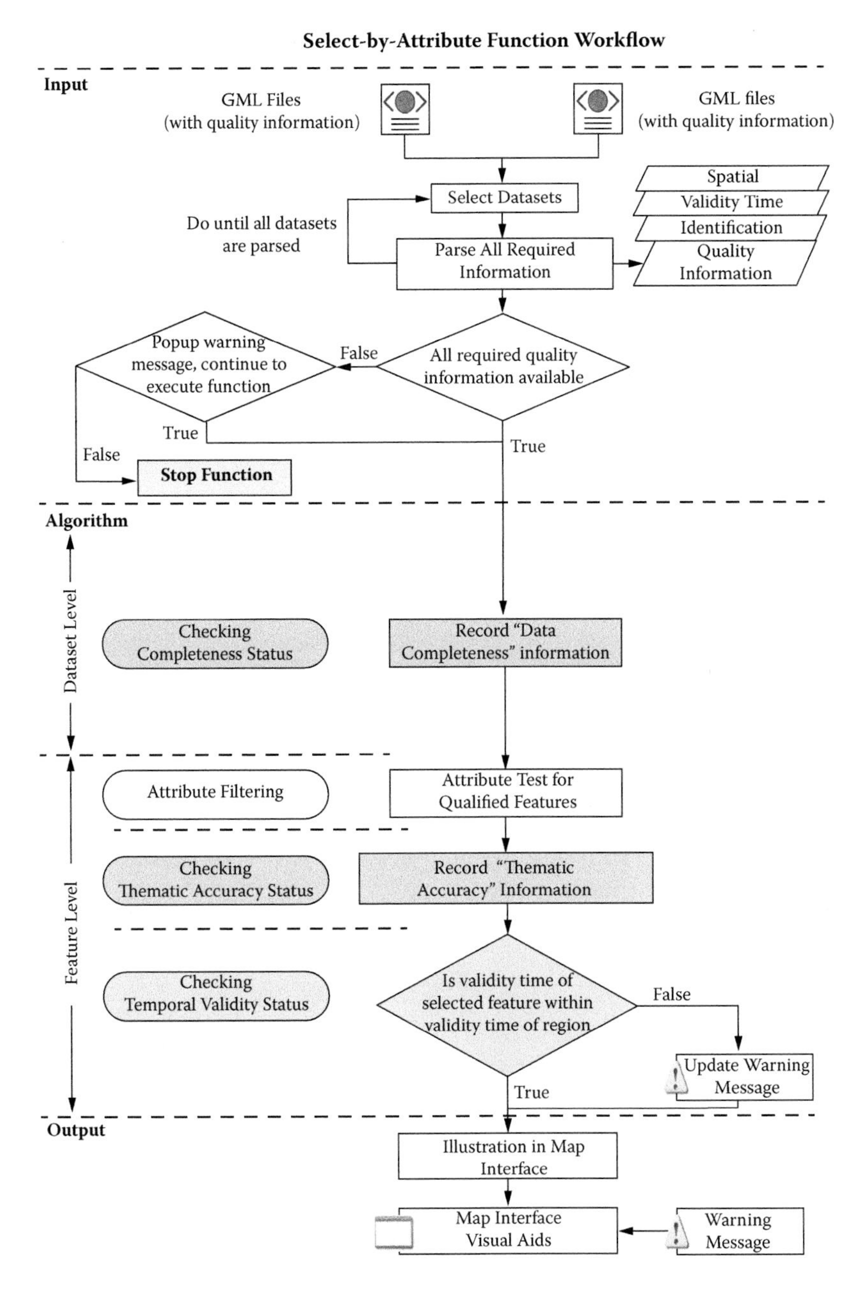

FIGURE 4.5
Workflow of the select-by-attribute function.

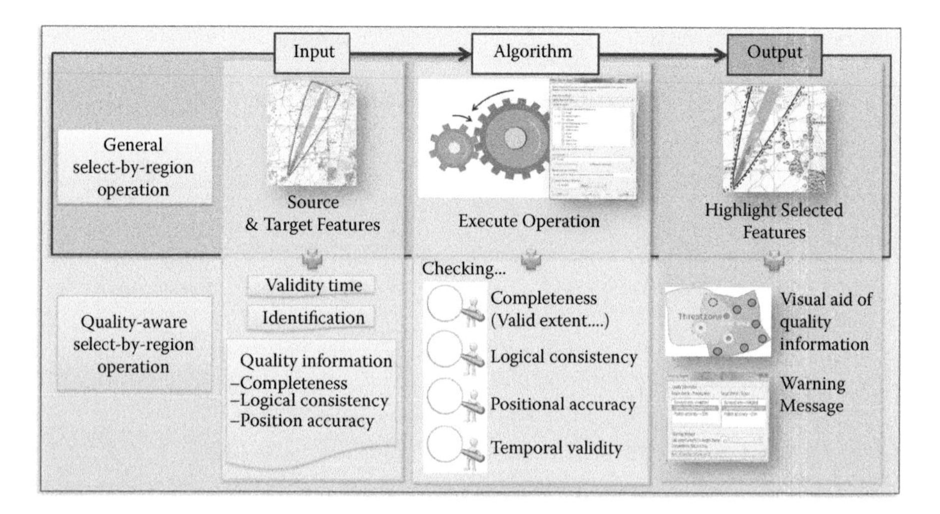

FIGURE 4.6
Workflow design of quality-aware GIS functions.

FIGURE 4.7
Simulated threat zone of chlorine gas exposures.

Theoretically, the evacuation decision is based on the topological relationship between the features and the threat zone, that is, whether the features are *within* the threat zone. However, the result is correct only when the surveyed area of the selected data set *contains* the spatial extent of the threat zone. Otherwise, only parts of features for the chosen themes are possibly found, and thus the decision may be different. Warning messages or visual aids must be prompted to users to inform them of the possible risks. As this case is often an emergency situation, decisions based on incomplete or outdated data may potentially lead to serious public damage. Even in cases where the surveyed area contains the spatial extent of the threat zone, if data are incomplete, users must be warned that certain features may be missing from the selected results.

Figure 4.8 illustrates the selection results of the traditional select-by-region function. Through the map window, users can visually identify the location of selected features and later use other functions to calculate the number of features, create statistical reports on affected citizens, and prepare an evacuation plan. Although the preceding discussion seems highly reasonable, Figure 4.9 illustrates the results after applying the modified select-by-region function. The spatial extent of the threat zone is subdivided into two major categories. The polygon depicted by oblique grid symbols represents the overlapped region of the predicted chlorine gas threat zone and the surveyed area of the building data set, whereas the other polygon represents an area where no building information is available for analysis. Although buildings within the overlapped area can be found, no information about the nonoverlapped area is available. Further superimposition of the surveyed

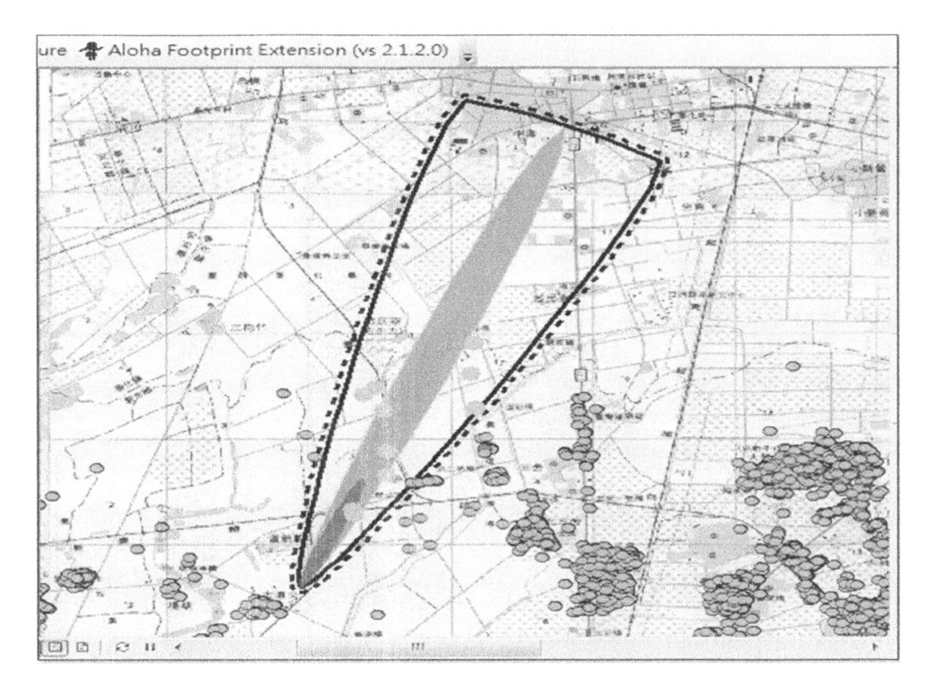

FIGURE 4.8
Result of using a traditional select-by-region function.

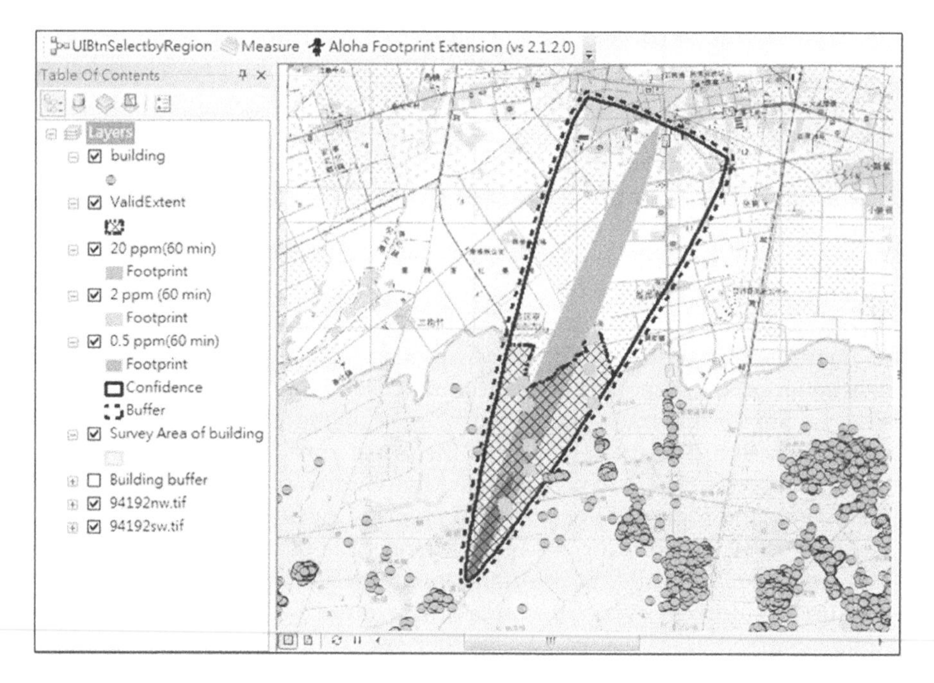

FIGURE 4.9
Result of using a quality-aware select-by-region function.

area of the building data set (yellow polygon) in the map window shows a more obvious scenario of missing information. Generation of the polygon of oblique grid symbols is extended from the concept of valid extent proposed by Hong and Liao (2011). By assuming the threat zone to be a data layer, we can use the geometric intersection approach to deduce the area for which quality can be evaluated, and we can also identify the high-risk area for decision making. For example, if the rate of omission error for the building data set is 2%, then believing that some buildings are missing from the surveyed area is reasonable.

In addition to the analysis of data completeness, the parsed quality information is automatically prompted to users for further analysis or reference (Figure 4.10).

The preceding discussion shows that even the widely used select-by-region function requires a new design. The traditional function design deals only with the features present in the data set. Although this design can find features that meet specified constraints, it fails to determine whether all of the features have been found. In this case, the newly introduced concept of surveyed area and the corresponding valid extent algorithm can help to provide meaningful interpretations, point out the areas of risk, and prompt with visual aids to capture the attention of the users.

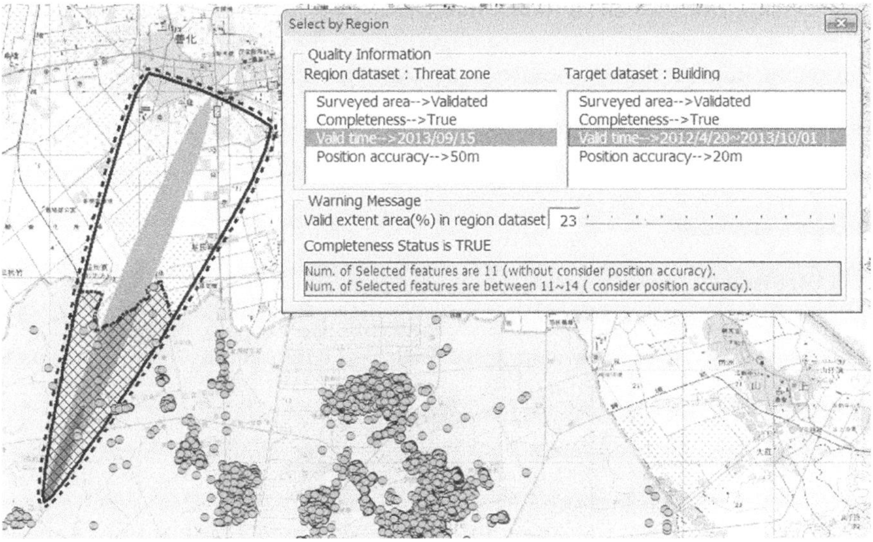

FIGURE 4.10
Design of visual aids for quality information.

4.5 Conclusion and Future Outlook

The development of SDI facilitates a powerful data-sharing mechanism for various types of users to take advantage of versatile georesources distributed over the Internet. Because users frequently have to deal with unfamiliar or even unknown data, the role of data quality information should not be restricted to only auxiliary information for finding data. Instead, correct use of data in the applications should be a mandatory consideration for users. We argue that the design of GIS functions must be reexamined to include the consideration of data quality so that users become automatically aware of the quality status of the results. We demonstrate that even a simple and straightforward function may require multiple data quality components and knowledge-driven algorithms to ensure the correct interpretation of results. A standardized link between geospatial data and standardized metadata is proposed so that GIS functions can accordingly parse necessary quality information for further analysis. The GIS interface must have a built-in mechanism to unambiguously explain, graphically or otherwise, the quality of the processing results. The standardized encoding, knowledge-driven algorithm, and quality-aware visual aid design bring a new and enhanced paradigm to future GIS design. To address the increasingly complicated challenges of integrating different data resources into future web-based and cloud computing environments, the innovative integration of quality-aware GIS and OpenGIS enables an intelligent and interoperable application environment in which even novice users can safely make decisions based on what is illustrated on the map interface.

References

Devillers, R., Bédard, Y., Gervais, M., Jeansoulin, R., Pinet, F., Schneider, M., and Zargar, A. 2007a. How to improve geospatial data usability: From metadata to quality-aware GIS community. Presented at Spatial Data Usability Workshop, AGILE 2007 Conference, Aalborg, Denmark.

Devillers, R., Bédard, Y., and Jeansoulin, R. 2005. Multidimensional management of geospatial data quality information for its dynamic use within GIS. *Photogrammetric Engineering and Remote Sensing*, 71(2): 205–215.

Devillers, R., Bédard, Y., Jeansoulin, R., and Moulin, B. 2007b. Towards spatial data quality information analysis tools for experts assessing the fitness for use of spatial data. *International Journal of Geographic Information Science*, 21(3): 261–282.

Devillers, R., Stein, A., Bédard, Y., Chrisman, N., Fisher, P., and Shi, W. 2010. Thirty years of research on spatial data quality: Achievements, failures, and opportunities. *Transactions in GIS*, 14: 387–400.

Devillers, R., and Zargar, A. 2009. Towards quality-aware GIS: Operation-based retrieval of spatial data quality information. Presented at Spatial Knowledge and Information (SKI) Conference, Fernie, British Columbia, Canada.

Dustdar, S., Pichler, R., Savenkov, V., and Truong, H. 2012. Quality-aware service-oriented data integration: Requirements, state of the art and open challenges. *SIGMOD Record*, 41(1): 11–19.

Goodchild, M.F. 2004. A general framework for error analysis in measure-based GIS. *Journal of Geographical Systems*, 5(4): 323–324.

Hong, J.-H., and Liao, H.-P. 2011. Incorporating visualized data completeness information in an open and interoperable GIS map interface. *Journal of the Chinese Institute of Engineers*, 34(6): 733–745.

International Organization for Standardization (ISO). 2002. Geographic information—Quality principles. ISO 19113. ISO, Geneva.

International Organization for Standardization (ISO). 2003a. Geographic information—Quality evaluation procedures. ISO 19114. ISO, Geneva.

International Organization for Standardization (ISO). 2003b. Geographic information—Metadata. ISO 19115. ISO, Geneva.

International Organization for Standardization (ISO). 2006. Geographic information—Data quality measures. ISO 19138. ISO, Geneva.

Lanter, D.P., and Veregin, H. 1992. A research paradigm for propagating error in layer-based GIS. *Photogrammetric Engineering and Remote Sensing*, 58(6): 825–833.

Maguire, D., and Longley, P. 2005. The emergence of geoportals and their role in spatial data infrastructures. *Computers, Environment and Urban Systems*, 29(6): 3–14.

U.S. Environmental Protection Agency (USEPA) and the National Oceanic and Atmospheric Administration (NOAA). 1999. *Area Locations of Hazardous Atmospheres (ALOHA), User's Manual*. USEPA and NOAA.

Yang, T. 2007. *Visualisation of Spatial Data Quality for Distributed GIS*. University of New South Wales, Australia.

Zargar, A., and Devillers, R. 2009. An operation-based communication of spatial data quality. In *2009 International Conference on Advanced Geographic Information Systems and Web Services*, pp. 140–145, Cancun, Mexico.

Section II

Uncertainties in Multidimensional and Multiscale Data Integration

5

Data Quality in the Integration and Analysis of Data from Multiple Sources: Some Research Challenges

J. Harding, L. Diamond, J. Goodwin, G. Hart, D. Holland, M. Pendlington, and A. Radburn

Ordnance Survey, Southampton, United Kingdom

CONTENTS

ABSTRACT This chapter describes preliminary work to investigate what it means to manage data quality in a simple data integration and analysis prototype for research purposes, where input data sets are from a range of different sources. Consideration is given to how standard elements of spatial data quality (as in ISO 19115:2003) apply in the context of the prototype, which is based on a relatively straightforward house-hunting scenario. Based on initial findings, the chapter aims to position further work, identifying a series of research questions around needs for improved data quality management and communication in analytical processes involving geographic information. It is hoped the chapter may serve to add support for more applied and user-focused data quality research in the area of analytics.

5.1 Introduction

Many applications of geographic information involve some kind of decision making or deriving insight based on the information available. From viewing a map for trip planning to carrying out complex spatial analyses of geographic data, a fundamental consideration is, how fit are the data for the specific use and user? Often the problem or decision requires data from different sources to be used together. Within navigation systems, for instance, route network, place-name, and address data may be integrated with live traffic information. Using such systems, a decision on route choice is likely to be more critical for an ambulance driver than for a tourist, and the consequences more severe if the combined routing information is wrong in some way. Providing information about data quality that is relevant to use context and communicated in a meaningful way can allow users to assess reliability and fitness for their purpose.

Moving beyond established GIS applications to uses of geographic information in what is loosely referred to as Big Data analytics and other analysis where geographic information is not necessarily the central focus, end users are interested in answers to queries or insights from data but might not themselves be involved in the data processing. In this way, users of analytical output may be even more removed from the source data but could still find information concerning data quality helpful to assess reliability of analytical output, thereby enabling them to reduce uncertainty or risk in their use of the output. Developments to realize new value and insight from data through Big Data analytics and business intelligence systems, involving data integration from multiple sources, data mining, and visualizations of outputs, for example, present questions around quality of inputs and outputs and how significant are aspects of data quality for a given use context.

System developers are recognizing that data quality processes are important to success in Big Data analytics, and with location being increasingly seen as a valuable dimension within business intelligence systems, there is growing focus on how geospatial standards can be utilized to enhance these systems (e.g., OGC, 2012). While a survey of industry managers and analysts suggested that data quality was not perceived as a major obstacle to the uptake of business analytics when compared to managerial and cultural factors (LaValle et al., 2011), it remains an area about which increased awareness is needed to inform information use and reduce risk of misuse (Gervais et al., 2009). For providers of analytical services (where data delivered to client users are in the form of query and analysis results), this means being able to understand and manage data quality starting with input data, through the analytical processes, to the output data quality and presentation of that data.

This chapter describes preliminary work to investigate what it means to manage data quality in a simple data integration and analysis research prototype, where input data sets are of disparate types and from a range of open data sources. Consideration is given to standard elements of spatial data quality (as in ISO 19115:2003) and how they apply in the context of the prototype, which is based on a relatively straightforward house-hunting scenario. While not a critical application, misleading output information would be of little value to users and might undermine trust. Based on initial findings, the chapter aims to position further work, identifying a series of research questions around needs for improved data quality management and communication in analytical processes involving geographic information.

5.2 Research Approach

5.2.1 Data Processing Aims

An overall aim of the research prototype was to demonstrate the feasibility of using linked data approaches to integrate georeferenced data of different types from different sources and to use a combination of NoSQL and Resource Description Framework (RDF) database technologies to store large volumes of data and retrieve spatial query results efficiently via a web-based user interface. The benefits of these technologies include their scalability and flexibility to handle evolving content in both structured and unstructured formats.

5.2.2 User Scenario and Interface Design

Prototype development focused on a simple scenario of a family moving to a new area and needing information about different neighborhoods to help them focus their house hunting. Based on setting a range of criteria reflecting their priorities, the users would be presented with a list of unit postcodes (areas encompassing on average around 15 addresses) that meet these criteria within the limits of their geographical search area. For purposes of this research, with emphasis on technical proof of concept, the research project team identified likely criteria of interest in a family house-hunting scenario: proximity to schools and proximity to points of interest, including pubs and food outlets (for convenience or to avoid), doctors' (GP) surgeries, and supermarkets.

Figures 5.1 through 5.3 show screenshots from the user interface, enabling the users to first select the geographic area of interest and then select the relevant factors and value ranges for their search criteria and view results.

As an example, the users are interested in living within a 1-mile radius of the center of Oxford.

Step 1: Select area of interest. The users enter the place-name they wish to center the search on (Oxford) and the search radius. A map display with the required search area is returned, the system having used Ordnance Survey gazetteer data to locate the desired search location (Figure 5.1).

Step 2: Factors—what are you interested in? The users check boxes in the interface for factors affecting their search (e.g., proximity of schools and GP surgeries).

Step 3: Parameters—low–high, near–far. Using sliders representing the selected factors, the users set high and low limits (for those illustrated in Figure 5.2, these are distance ranges).

Step 4: Postcode results. The system retrieves a list of postcodes that meet the search criteria and displays their centroids on a map (Figure 5.3). By clicking on a postcode, the users can view names of features meeting their search criteria, for example, "Cherwell School" or "Royal Oak Pub." These details are retrieved through querying all data sources in the system that have content relevant to the criteria selected. An outline of the research prototype development is given in Section 5.2.3.

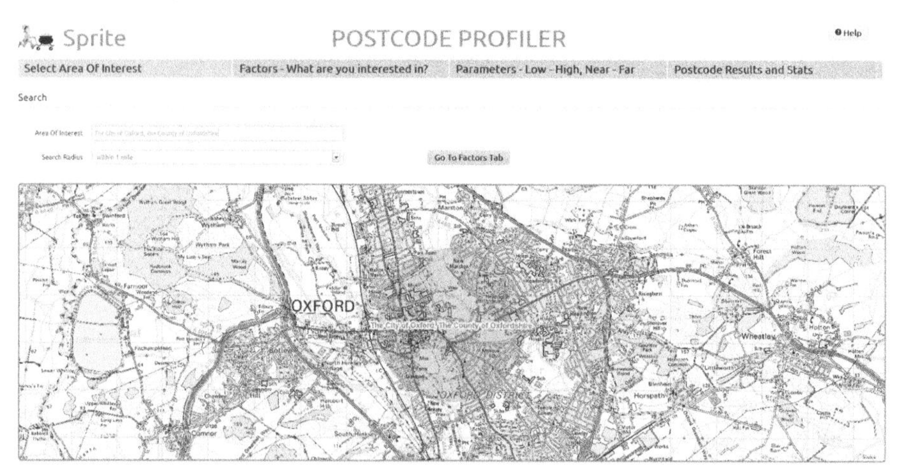

FIGURE 5.1 (See color insert.)
Home page of the Postcode Profiler interface for users to select their area of interest. (Sprite was the internal research project name.) (Crown® Copyright and database right, 2014. All rights reserved.)

Sprite

POSTCODE PROFILER

Help

Select Area Of Interest | Factors - What are you interested in? | Parameters - Low - High, Near - Far | Postcode Results and Stats

Get The Results

Crime	0 to 99 (0 = highest crime, 99 = lowest crime)		
High		Low	
Primary Schools	0.1 to 0.5 Km		Set far value (in Km)
Near		Far	
Secondary Schools	0.1 to 1.0 Km		Set far value (in Km)
Near		Far	4.0
GP Surgeries	0.1 to 0.3 Km		Set far value (in Km)
Near		Far	
Pubs, Bars and Inns	0.5 to 1.0 Km		Set far value (in Km)
Near		Far	
Takeaway Outlets	0.5 to 1.0 Km		Set far value (in Km)
Near		Far	
Supermarkets	1.0 to 5.0 Km		Set far value (in Km)
Near		Far	

FIGURE 5.2
Third tab of the Postcode Profiler interface, enabling the users to define parameter values for querying the linked data store, having previously selected factors of interest from a wider range of options. (Crown® Copyright and database right, 2014. All rights reserved.)

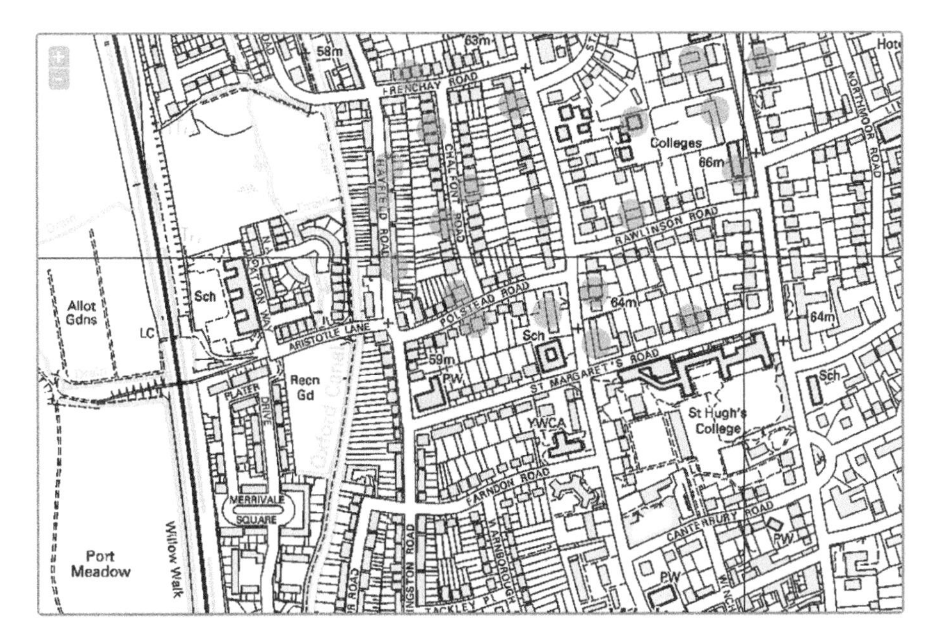

FIGURE 5.3 (See color insert.)
Extract from research results page showing centroids of postcodes that meet the users' search criteria. (Crown® Copyright and database right, 2014. All rights reserved.)

5.2.3 Prototype Development

In outline, the development process involved the following:

- Acquiring data sets of third-party data relevant to the selected user scenario, together with relevant data sets from the Ordnance Survey in Great Britain (see Table 5.1).
- Converting the third-party data, which were from a variety of sources, to a unified .csv format (if not already in .csv) using Python scripts.
- Converting the .csv data to a unified model of linked data "triples," consisting of expressions made up of named nodes and the relationship between them (written using RDF) (W3C, 2004) to establish linkages between the data sets. Connections between data sets were made using postcode uniform resource identifiers (URIs).
- Loading to a "triple store" for storage (the open-source Apache Jena TDB was used). Each data set was stored in a named graph, allowing for efficient versioning and update. Metadata about each data set were also stored in RDF form. Spatial data were stored in MongoDB (an open-source document database) for spatial query processing efficiency.
- Communicating between components using web services.
- Providing an interface enabling user interaction through the setting of search criteria that drive SPARQL queries across the data, as well as presenting results returned in response to these user-selected criteria.

TABLE 5.1

Data Required for the Prototype and Sources Used (2012)

Data Type Required	Source/Provider	Dataset
House prices	HM Land Registry, data.gov.uk	Price Paid Data[a]
Deprivation levels	Office for National Statistics, data.gov.uk	Index of Multiple Deprivation[a]
Crime statistics	Home Office (publisher), data.police.uk	Street-level crime data
Primary and secondary school locations	Department of Education, data.gov.uk	EduBase, Schools in England[a]
GP surgery locations	HSIC Organisation Data Service, data.gov.uk	GP practices[a]
Points of interest (various)	PointX	Points of Interest[b]
Zoomable backdrop map and gazetteer data to enable place-based search	Ordnance Survey	OS OpenData[a] products

[a] Contains public sector information licensed under the Open Government Licence v2.0.
[b] This product includes data licensed from PointX (copyright Database Right, 2012).

5.2.4 Quality of Data Inputs and Outputs

Credibility of the output results would in part depend on qualities of the various source data and any data transformation in the creation of linked data or in the analysis process invoked by a user-defined query. It was therefore important to understand as far as possible the quality of source data and any changes to data quality in the data transformation processes, and be able to communicate information about the quality of output information in a meaningful way to the users given the type of query or analysis enabled through the user interface. In this way, both internal data quality of source data, as provided by the data set creators, and external data quality of outputs, as communicated to the end users through the interface of the prototype, were within scope of this study.

5.2.5 Data Quality Elements Relevant to the Scenario

Qualitative assumptions were made, as follows, about minimum levels of data quality that matter in the use context:

Positional accuracy: For the schools and other points of interest, an address point location would be sufficient to determine presence within a postcode. For other criteria, for example, values for house prices, crime rates, or deprivation levels, a value or range of values applicable at postcode resolution would be sufficient. In this way, spatial granularity of data was more significant for some of the required data elements than positional accuracy of individual data points; that is, the positional information of much of the attribute data needed to be related to postcodes, areas of coarser spatial granularity than the more precise (to the nearest meter) position of an address.

Attribute accuracy: It was important that all criteria were represented by attribute values true to reality to a level of detail appropriate to the scenario. For example, in the case of schools, attribution as primary or secondary education needed to be correctly applied.

Temporal validity: For all criteria the most up-to-date data were required. Actual creation or capture date and update schedules were expected to be different between data sets, so as a minimum requirement, the creation date of the data set and assurance that the data were of the latest data release needed to be known. Data for house prices, for example, might be misleading if more than 2 years old, whereas levels of deprivation are subject to slower change and older data (as long as they are the most recently published) may be still relevant. Data for some criteria, such as house prices and crime statistics, might relate to a specified time period, in which case the bounding dates to which they apply need to be known.

Logical consistency: Each data set needed to be logically consistent within itself and according to its data model or specification in order to facilitate translation to the linked data format (RDF) used in the prototype. Even if used without translation to RDF, logical consistency errors would impede loading to a database and running data queries or analyses.

Lineage or provenance: At a minimum for the user scenario, knowing where the data have come from in terms of source or creating organization can help provide a basis on which the users can judge their level of trust in the output information. Understanding how that data were created is probably unnecessary in this particular scenario, but data transformations that happen within the prototype application could improve or degrade source data quality and may be significant enough to communicate to the end user. Recording lineage within the data process is therefore an important internal consideration that may require some level of description alongside data that are output by the prototype.

Completeness: For all criteria it is important to know that query output is based on complete data, meaning that postcodes not listed in query output are absent because they genuinely do not meet the selected criteria, rather than because there are data omission errors in source datasets. It is therefore important internally to the prototype that source datasets are complete with respect to their specification, or at least to recognize where datasets may not be expected to be complete. For example, if data were captured from voluntary sources, they cannot be expected to be complete.

Further considerations concerned coverage (geographic coverage for England was needed) and georeferencing system used. An internal requirement of the prototype was for all source data to be georeferenced in some way (e.g., by National Grid coordinate, latitude and longitude, or postcode) or to have unique identifiers enabling linkage to a georeferenced source (e.g., a Unique Property Reference Number [UPRN]).

5.3 Results

5.3.1 Data Sources

For purposes of the proof of concept, open government data sets were sought with content potentially suitable to serve the house-hunting scenario at the assumed minimum quality levels. The data sets used were mostly sourced from www.data.gov.uk but also included a commercial points of interest data set. In addition, Ordnance Survey OpenData and commercial products were used in order to present results in the user interface against a zoomable map backdrop.

5.3.2 Input Data Quality and Uncertainties

Generally, across data-set sources used, the availability of metadata for data quality was very limited, either within the data set or in associated documentation. For each data source, there was no single document or web page that had all the information sought. The following summarizes uncertainties about levels of quality with respect to elements listed in Section 5.2.3.

Positional accuracy: In most cases, the type of georeferencing used in a data set (e.g., postal address or postcode centroid) indicated the degree of positional accuracy or granularity of data to be expected. In all cases, however, correct positioning of data could not be verified unless it were to be compared with alternative sources of position for the same feature. Alternative sources would need to be of different provenance to afford an independent comparison. This was not undertaken within the prototype.

Attribute accuracy: Likewise, correct attribution could not be verified unless it were to be compared with alternative sources of the attribute for the same feature. This was not undertaken within the prototype.

Temporal validity: For most sources, metadata were provided about the date range that the data set applied to. Some also provided a data set creation date or publication date. Uncertainty remained over whether the data set was the most up to date available unless information about update schedule was also provided.

Logical consistency: Few errors were found in logical consistency during the conversion of source data to RDF. This conversion process was itself a way of validating logical consistency where required. Errors that did occur in source data were due to postcode syntax (e.g., S016 0AS instead of SO16 0AS).

Lineage or provenance: For all data sets, provenance in terms of source organization or publisher name was directly obtainable, together with license terms and, in some cases, advice on attribution statements to be used. Information about lineage in terms of the data creation process was in most cases less accessible. For some data sets, however, for example, the Index of Multiple Deprivation, the data creation process can be found in separate technical documentation (CLG, 2011).

Completeness: Where data sets in effect provide complete coverage of a choropleth type, a contiguous coverage of polygons would be expected with a value per attribute for each polygon. Therefore, for these kinds of data, errors of omission or commission should be detectable on ingesting the data. Only the Index of Multiple Deprivation was of this type. For the other data sets involved, completeness could not be verified.

Coverage: All the sources used provided information on geographic coverage for the data set, usually by country name rather than geometry of extent.

5.3.3 Data Processing Causing Change to Data Quality

Source data were changed in terms of required fields being converted to linked data, where data were not already in this format. In the case of logical consistency errors, this conversion process resulted in removal of errors and thereby improvement to logical consistency. Such change due to data processing becomes part of data lineage within the prototype. A record of this process was produced for one of the input data sets, HM Land Registry house price data, using World Wide Web Consortium (W3C) provenance vocabularies (W3C, 2013a), in order to demonstrate the potential for managing lineage in linked data structures.

5.3.4 Communicating Quality of Output Query Results

Given the needs of the user scenario outlined above and uncertainties in many elements of source data quality, it was decided in this prototype to focus on those elements of most certainty and relevance to the user scenario. These were the provenance and temporal validity of the data sources. Search results returned in the user interface were accompanied by a table showing data sources used to provide the results, date of creation or publication, and date range of the data, if applicable. These metadata were retrieved from the RDF data-set metadata store.

5.4 Discussion

5.4.1 Summary of Experience from the Prototype Study

By reporting data provenance and relevant dates with the results of user-defined queries, the users of the prototype in this study at least have some basis for making their own judgment as to whether the information presented is a suitable aid to their house hunting. Nevertheless, in terms of providing a service from acquiring source data through to delivering query or analytical results, uncertainties exist in many of the elements of source data quality, as described above. In more rigorous or critical analytical scenarios, improved certainty about data quality may be necessary to enable fit-for-purpose outputs and enable user evaluation of risk. What can be done to reduce these uncertainties?

5.4.2 Source Data Quality as Input

We probably have to accept that creators of potentially useful data cannot all be relied upon to provide quality metadata in accordance with standards such as ISO 19115. Lack of complete metadata is a common issue, as highlighted by geospatial data experts surveyed on their approach to data-set selection (Lush et al., 2012), particularly in terms of provenance, lineage, and licensing information. In addition, they found that recommendations from within the user community, data provider reputation, and data provider comments on uncertainty and error estimates within their data influenced these specialist users' perceptions of quality. As found in the present study, some additional aspects of quality metadata not included in current standards would be helpful to know about source data, namely, the resolution or granularity of the data, and data set update schedules or intervals. In terms of positional and attribute accuracy, in the absence of quality statements provided with source data, an independent means of verification could help identify levels of uncertainty within the data and provide a basis for representing and communicating uncertainty to the end users, when important to output data usability. Uncertainty is inherent in much geographic data (e.g., as discussed by Couclelis, 2003; Duckham et al., 2001) in terms of both position and application of classifications to real-world things, yet this aspect of data quality is not represented in standards for data quality (Goodchild, 2007).

5.4.3 Data Quality in Data Processing

In the case of missing quality metadata, systems for data integration and analysis need means of independently assessing some aspects of the data's quality when this is important to output data usability. Depending on the type of analysis to be carried out, uncertainty in the spatial, temporal, and thematic dimensions of data used may each have impacts on the results of analysis. Zargar and Devillers (2009) review research that has linked the relative importance of these uncertainty dimensions to types of GIS operations and show how the communication of data quality information can effectively be linked to users' applications of operations. Where quality metadata are available, Devillers et al. (2007) go further in proposing a tool based on a multidimensional cube of compiled data quality information to provide data experts with meaningful information about known spatial data quality to support the required analysis.

Reporting lineage information from data source to output results could be significant in some use contexts. The graph structure of RDF allows storage of provenance metadata using W3C provenance vocabularies (W3C, 2013a). Further prototype development is needed to test whether this is more advantageous than storing metadata separately in tables, for example, based on the

Data Catalog Vocabulary (DCAT) (W3C, 2013b). Also, it is important to consider implications of data processing lineage for presenting attribution statements relating to analytical outputs, as well as for original source data used.

5.4.4 Output Data Quality

Where available, most quality metadata associated with source data tend to be created and expressed in a producer-centric way (Goodchild, 2007) and do not necessarily assist potential users in selecting suitable data. Investigating this issue, an analysis of information collected from customer interviews and feedback emails (Boin and Hunter, 2007) found that metadata were often confusing to data consumers. Opinions on suitability of a data set were sometimes derived from actual data content and comparisons with other information or ground truth, rather than quality metadata from the data supplier. For users of just the outputs from analytical services, data quality information needs to be communicated alongside analytical outputs with respect to relevance in the use context.

5.4.5 Questions Arising from This Study

A number of research questions relating to data integration and analytics are put forward from this short study. These may not be new, and some will already be subjects of research elsewhere. The intention here, however, is to identify some priorities for improved data quality management and communication as part of analytical applications involving geographic information.

Verifying source data: Where metadata for source data quality are lacking or insufficient, how can source data be verified to identify areas and levels of uncertainty within the data? Further, can source data be automatically verified against source data specifications (e.g., for positional accuracy, attribute accuracy, and completeness)? In other words, how well does the data conform to its capture or creation specification? Can source data content and quality be automatically verified against other sources of data?

Handling uncertainty in data processing: How can uncertainty and vagueness in geographic data be handled in data integration and analysis between different data sets? How can linked data structures handle geographic data uncertainty?

Confidence levels and communicating data quality: How can confidence levels in *source data* quality be represented and communicated effectively for different use contexts? How can confidence levels in quality of *output data* (resulting from analyses of data integrated from different sources) be determined and communicated effectively for different use contexts? How can inherent uncertainties or vagueness in source or output data be represented to the users in order to inform their decision making? How can data quality information be communicated effectively in different types of use contexts—what matters, what language, and what type of visualization of quality information are meaningful to the users in order to help them assess the risk of data use in decision making?

5.5 Conclusion

Preliminary work to investigate what it means to manage data quality in a simple data integration and analysis prototype has been explored in this chapter. Standard elements of spatial data quality (as in ISO 19115:2003) provide a useful basis for considering what elements of data quality are significant in a particular use context and for identifying the presence or absence of quality metadata associated with source data, but could usefully be extended to include factors of data granularity, uncertainty in data (spatial, temporal, and thematic), and the updating or releasing of schedules of data. These elements as categories all have relevance to the value of geographic information within developing Big Data analytics and business intelligence systems, involving data integration from multiple sources, analyses, and visualizations of outputs. With users of analytical output remote from the source data and not directly involved in the analytical operations applied, relevant data quality information presented in a meaningful way is needed to enable users to establish confidence or gauge risk in their use of the output. To this end, it is important for analytics service providers to understand as far as possible the quality of source data and any changes to data quality in data transformation and analysis processes, and be able to communicate information about the quality of output information in a way meaningful for the context of use.

The chapter aimed to position further work, identifying a series of research questions around verifying source data quality, handling uncertainty in data processing, and communicating meaningfully about the quality of output data. While not providing solutions, it is hoped the chapter may serve to add support for more applied and user-focused data quality research in the area of data analytics.

Acknowledgments

The authors gratefully acknowledge the work of other members of the Ordnance Survey research team involved in the project referred to in this chapter.

References

Boin, A.T., Hunter, G.J. 2007. What communicates quality to the spatial data consumer? Presented at Proceedings of the International Symposium on Spatial Data Quality 2007, Enschede, the Netherlands, June 13–15.

Couclelis, H. 2003. The certainty of uncertainty: GIS and the limits of geographic knowledge. *Transactions in GIS*, 7(2), 165–175.

Department for Communities and Local Government (CLG). 2011. The English Indices of Deprivation 2010. https://www.gov.uk/government/uploads/system/uploads/attachment_data/file/6320/1870718.pdf (accessed April 12, 2013).

Devillers, R., Bédard, Y., Jeansoulin, R., Moulin, B. 2007. Towards spatial data quality information analysis tools for experts assessing the fitness for use of spatial data. *International Journal of Geographical Information Science*, 21(3), 261–282.

Duckham, M., Mason, K., Stell, J., Worboys, M. 2001. A formal approach to imperfection in geographic information. *Computers Environment and Urban Systems*, 25, 89–103.

Gervais, M., Bédard, Y., Levesque, M.-A., Bernier, E., Devillers, R. 2009. Data quality issues and geographic knowledge discovery. In H. Miller and J. Han (eds.), *Geographic Data Mining and Knowledge Discovery*. CRC Press, Boca Raton, FL, pp. 99–111.

Goodchild, M. 2007. Beyond metadata: Towards user-centric description of data quality. Presented at International Symposium on Spatial Data Quality 2007, Enschede, The Netherlands, June 13–15.

International Organization for Standardization (ISO). 2003. ISO 19115:2003: Geographic information—Metadata. ISO, Geneva.

LaValle, S., Lesser, E., Shockley, R., Hopkins, M., Kruschwitz, N. 2011. Big data, analytics and the path from insights to value. *MIT Sloan Management*, 52(2).

Lush, V., Bastin, L., Lumsden, J. 2012. Geospatial data quality indicators. Presented at Proceedings of Accuracy 2012, Florianópolis, Brazil, July 10–13.

Open Geospatial Consortium (OGC). 2012. OGC white paper: Geospatial business intelligence (GeoBI). https://portal.opengeospatial.org/files/?artifact_id = 49321 (accessed April 19, 2013).

World Wide Web Consortium (W3C). 2004. Resource Description Framework (RDF). http://www.w3.org/RDF/ (accessed April 12, 2013).

World Wide Web Consortium (W3C). 2013a. Provenance Working Group. http://www.w3.org/2011/prov/wiki/Main_Page (accessed April 12, 2013).

World Wide Web Consortium (W3C). 2013b. Data Catalog Vocabulary (DCAT). http://www.w3.org/TR/2013/WD-vocab-dcat-20130312/ (accessed April 12, 2013).

Zargar, A., Devillers, R. 2009. An operation-based communication of spatial data quality. In *International Conference on Advanced Geographic Information Systems and Web Services*, pp. 140–145, Cancun, Mexico.

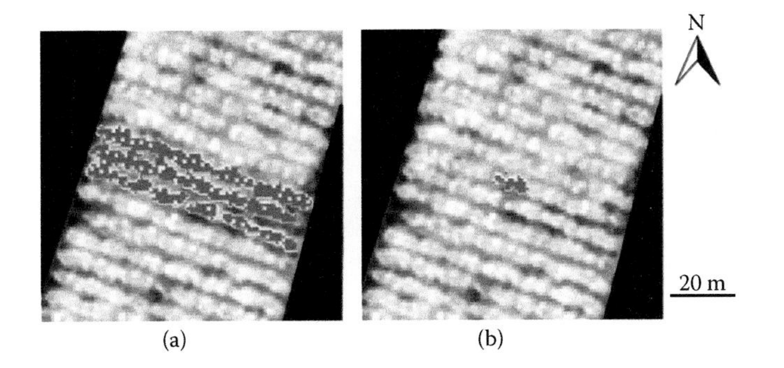

FIGURE 2.3
Combination of SRM with the actual ET (a) zooming in at three individual rows and (b) zooming in at three individual plants.

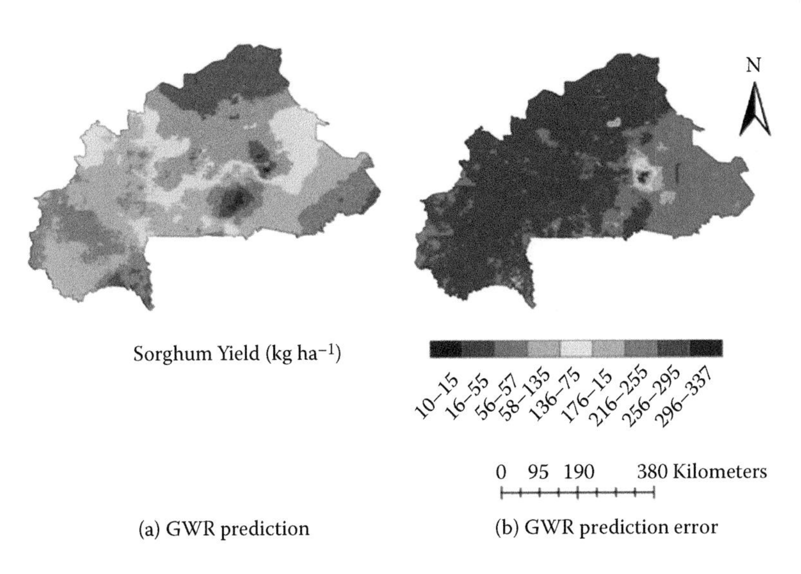

FIGURE 2.4
Predictions of sorghum yield at the national scale of Burkina Faso (left) and prediction errors (right). Both maps are able to show a large variation, due to the use of remote sensing images.

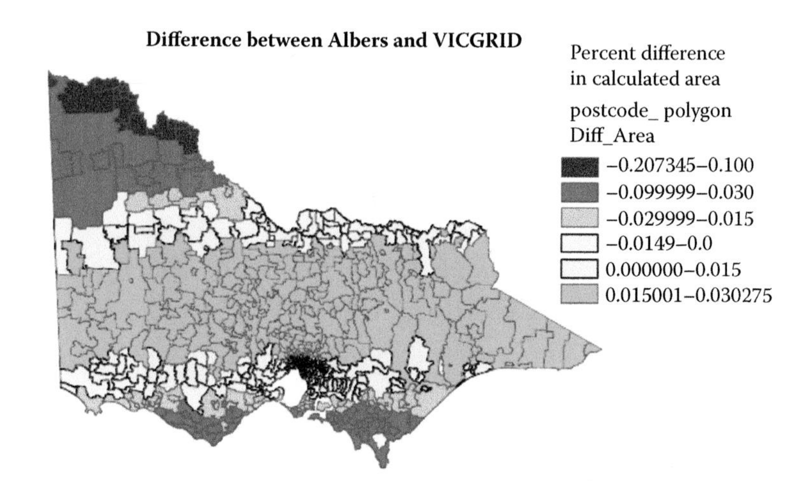

FIGURE 3.6
Difference between Albers and VICGRID in percent of area.

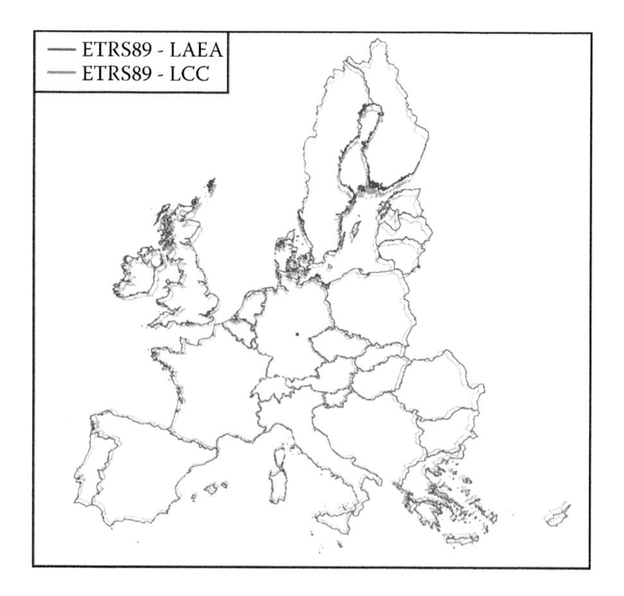

FIGURE 3.8
A representation of European Union countries using Lambert azimuthal equal-area (in blue) and Lambert conformal conic (in green) projections (inspired from Dana, 1997). (From http://epp.eurostat.ec.europa.eu/.)

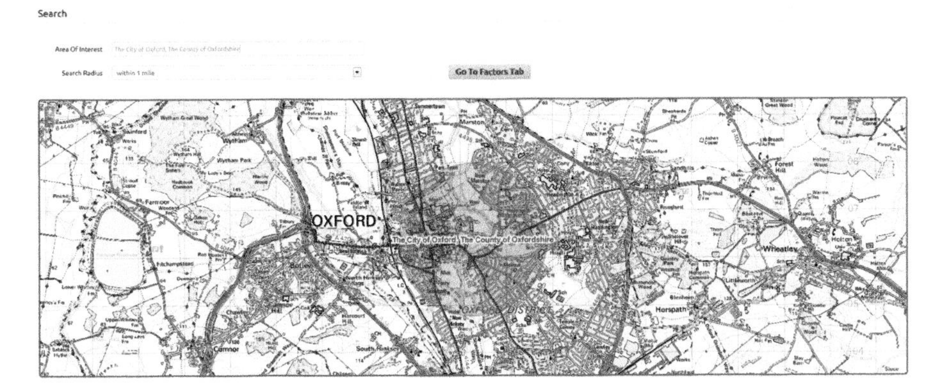

FIGURE 5.1
Home page of the Postcode Profiler interface for users to select their area of interest. (Sprite was the internal research project name.) (Crown® Copyright and database right, 2014. All rights reserved.)

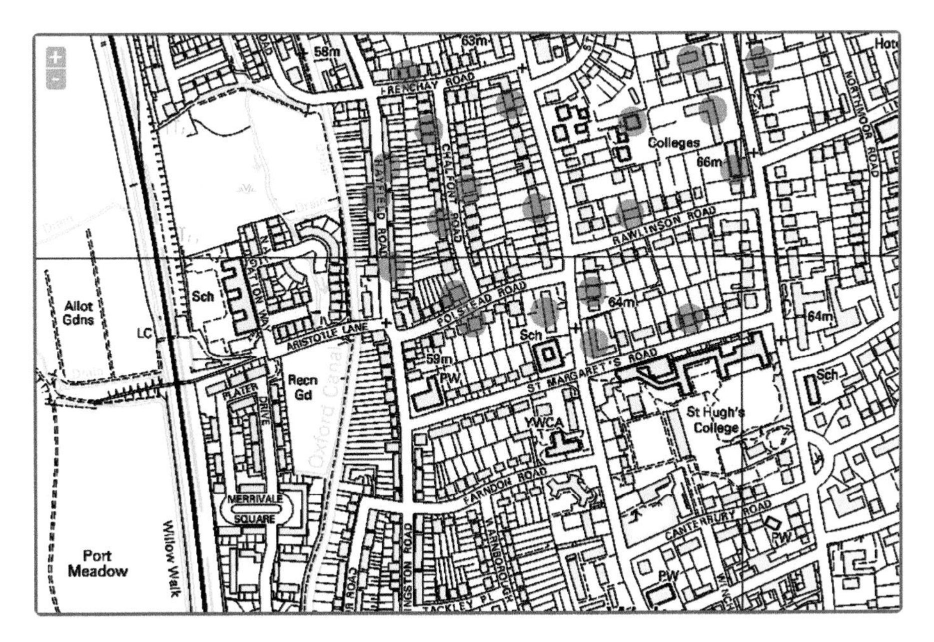

FIGURE 5.3
Extract from research results page showing centroids of postcodes that meet the users' search criteria. (Crown® Copyright and database right, 2014. All rights reserved.)

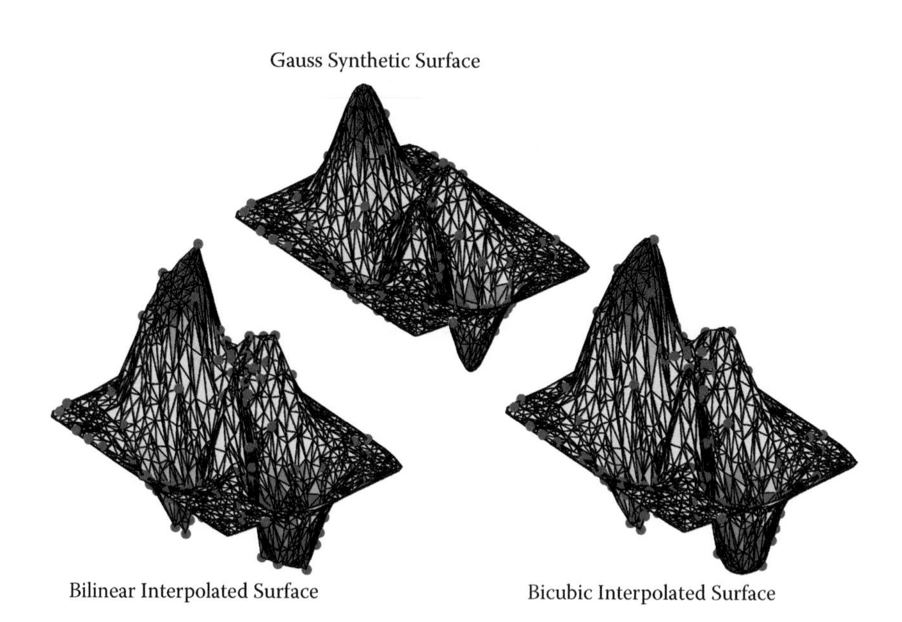

FIGURE 11.8
Test model error comparing the interpolated surfaces for TIN with the Gauss synthetic surface (m = 150, n = 1819).

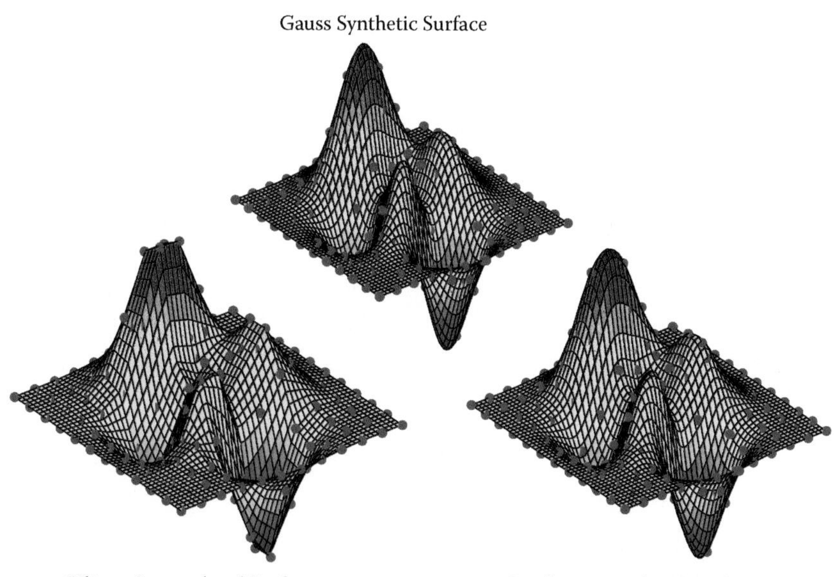

FIGURE 11.9
Test model error comparing the interpolated surfaces for rectangle with the Gauss synthetic surface (m = 144, n = 2401).

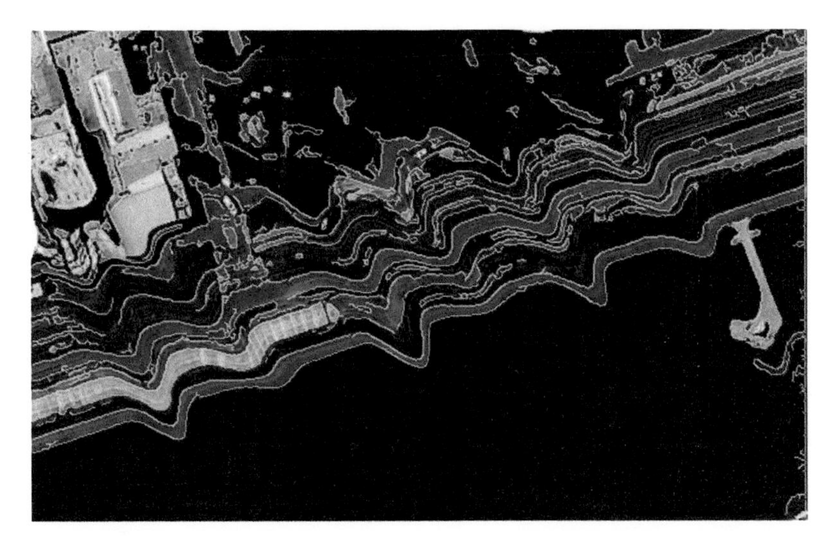

FIGURE 12.4
Contour lines after contour tracing.

FIGURE 12.5
Curve fitting results with cubic spline function.

FIGURE 12.6
The inspected squiggles in an aerial image.

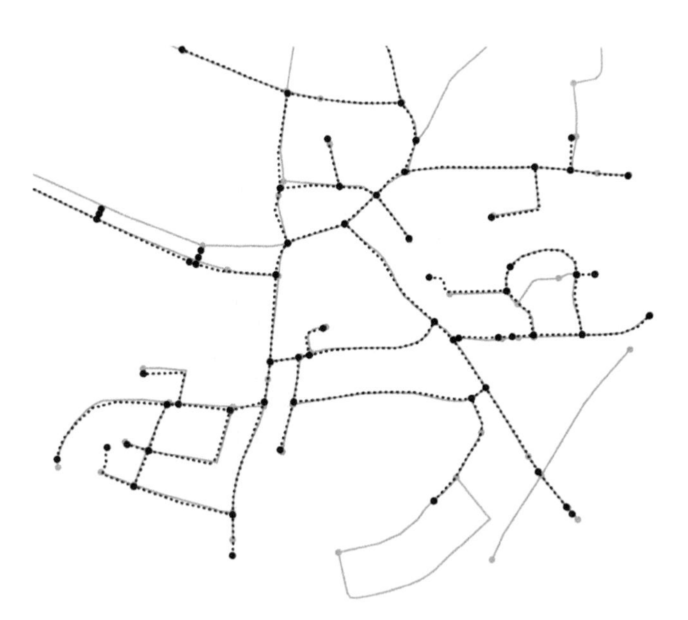

FIGURE 13.1
Overlay of two maps of the same region: gray solid, ATKIS; black dashed, OpenStreetMap.

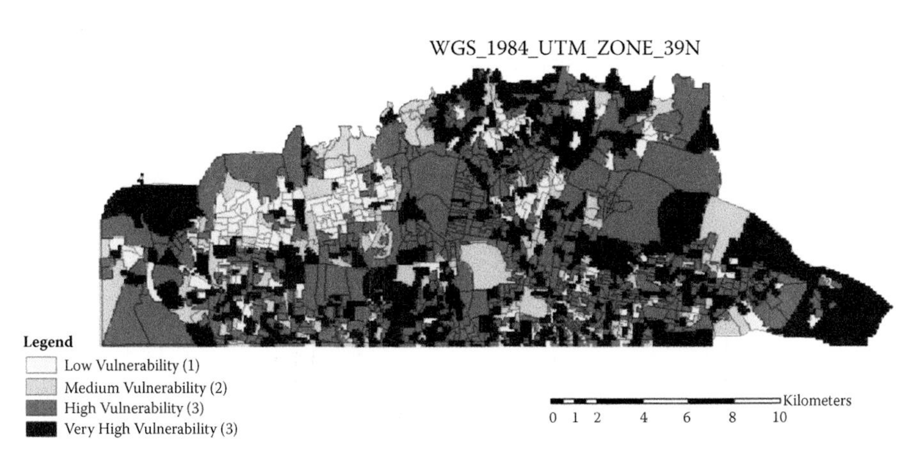

FIGURE 15.3
Seismic vulnerability map of the north of Tehran using the GrC model based on the neighborhood systems concept.

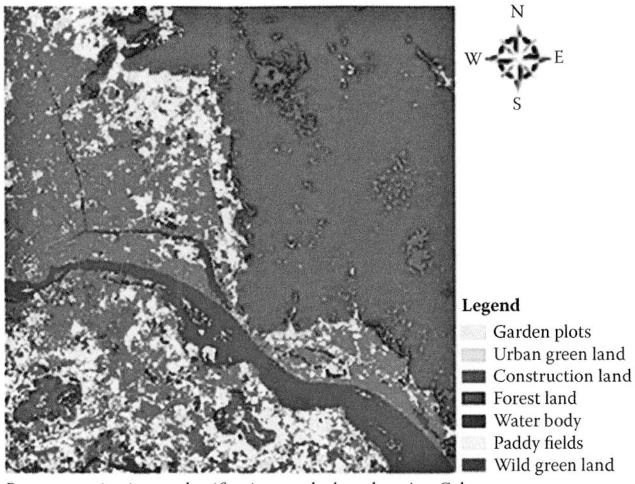

(a) Remote sensing image classification results based on Ant Colony Algorithm Optimization under the support of 8 variables

FIGURE 16.3
Classification based on ant colony algorithm for variables a–c. *(Continued)*

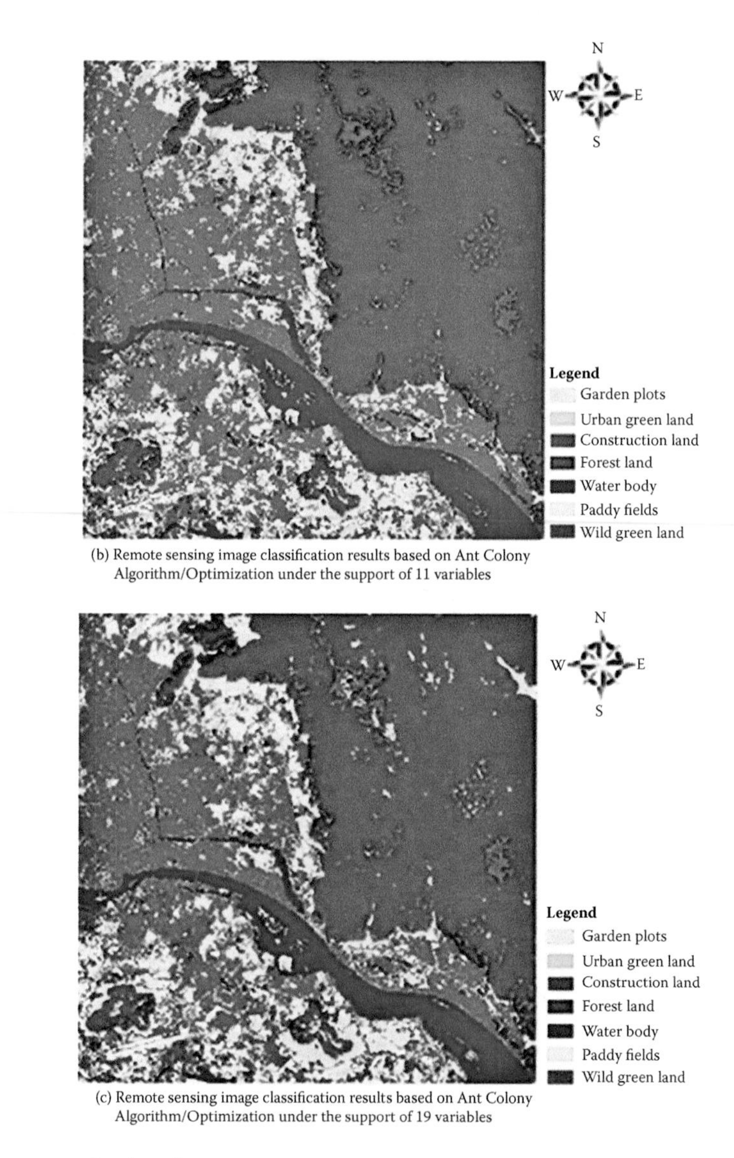

(b) Remote sensing image classification results based on Ant Colony Algorithm/Optimization under the support of 11 variables

(c) Remote sensing image classification results based on Ant Colony Algorithm/Optimization under the support of 19 variables

FIGURE 16.3 (Continued)
Classification based on ant colony algorithm for variables a–c.

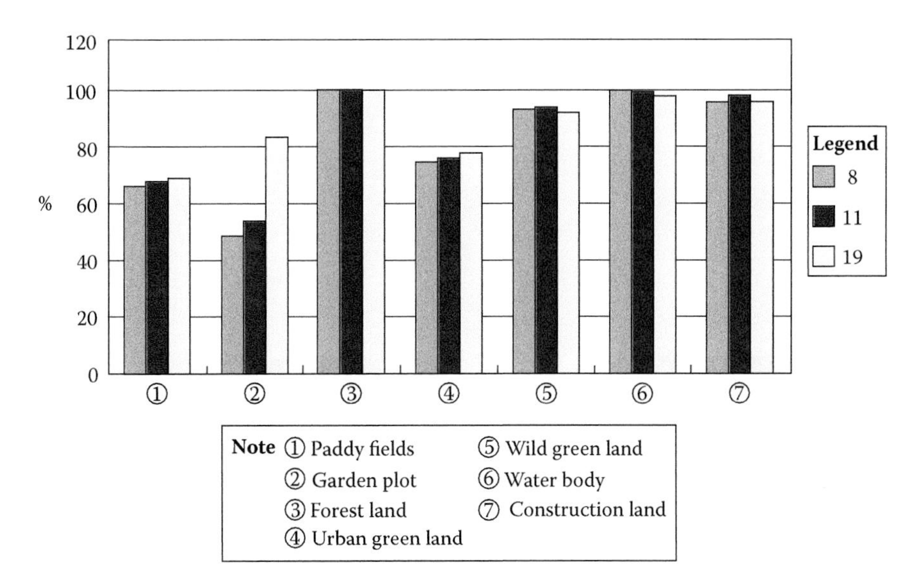

FIGURE 16.4
Comparison of producer accuracy based on 8, 11, and 19 variables.

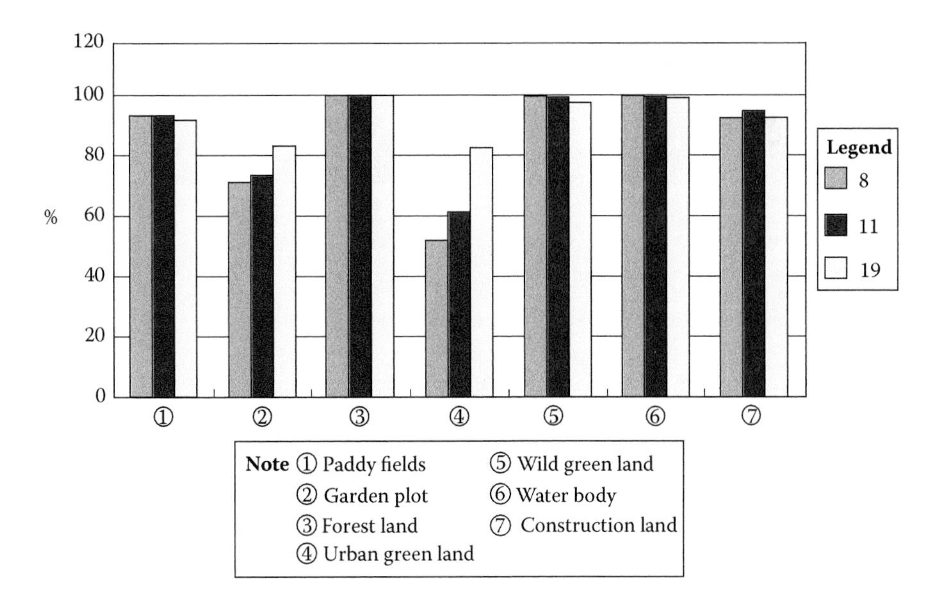

FIGURE 16.5
Comparison of user accuracy based on 8, 11, and 19 variables.

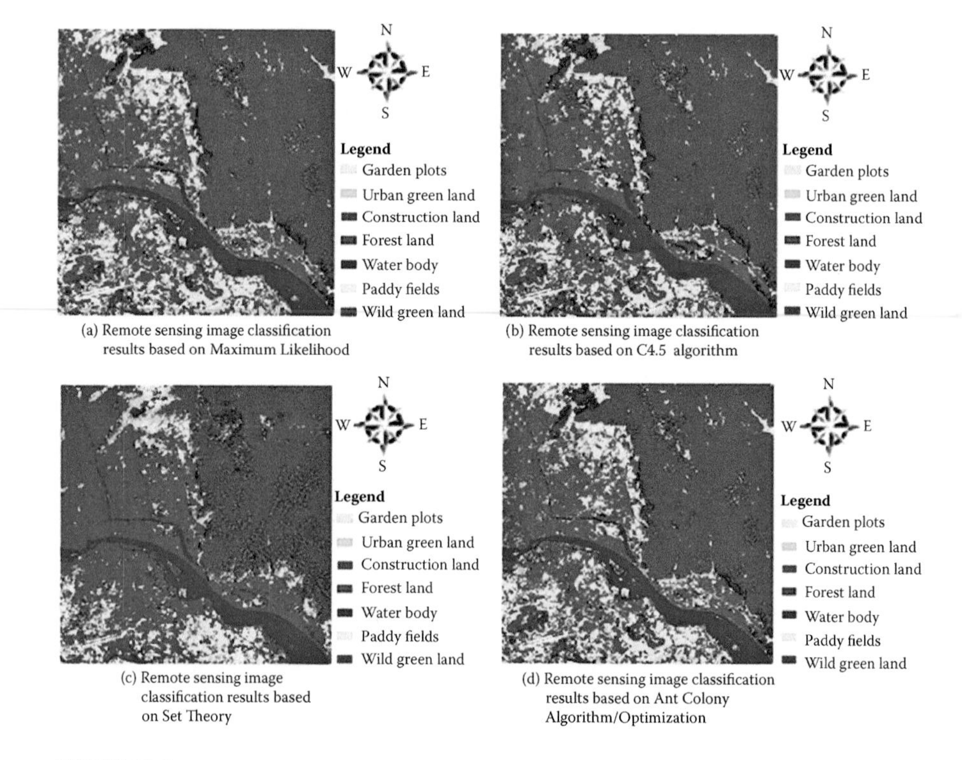

FIGURE 16.6
(a–d) Different classification methods based on remote sensing images.

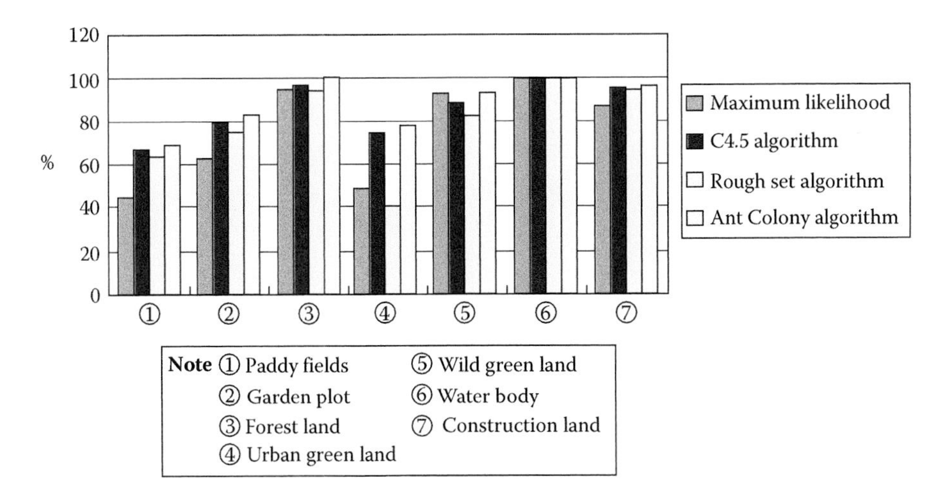

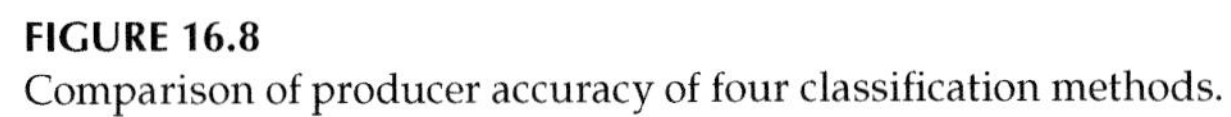

FIGURE 16.8
Comparison of producer accuracy of four classification methods.

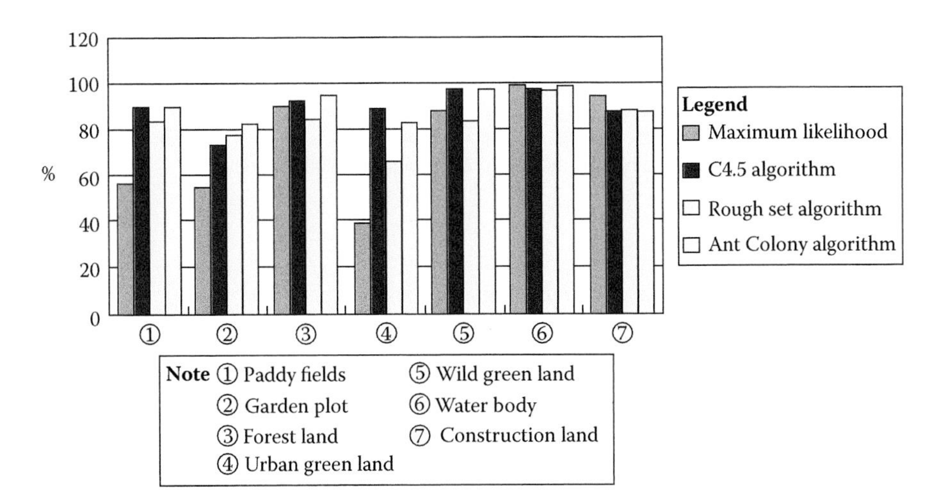

FIGURE 16.9
Comparison of user accuracy of four classification methods.

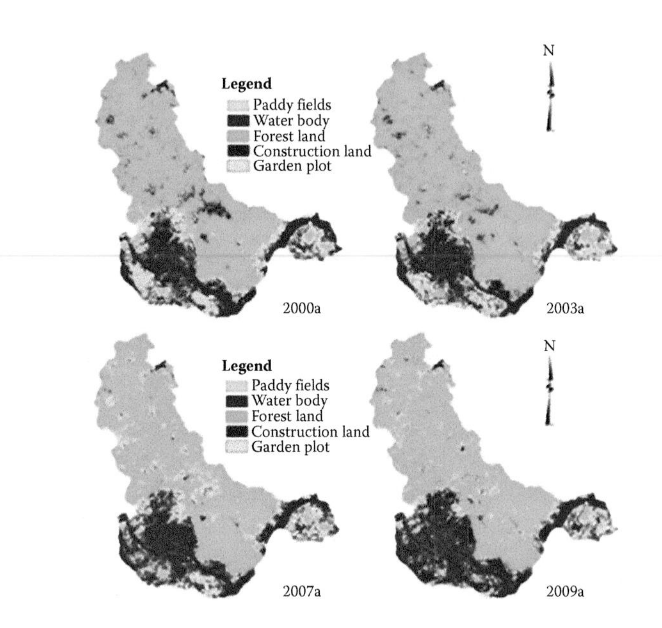

FIGURE 16.10
Classification images of Fuzhou from 2000 to 2003 and from 2007 to 2009.

6

Quality Management of Reference Geoinformation

A. Jakobsson,[1] A. Hopfstock,[2] M. Beare,[3] and R. Patrucco[4]

[1]*SDI Services, National Land Survey of Finland, Helsinki, Finland*

[2]*Federal Agency for Cartography and Geodesy (BKG), Frankfurt, Germany*

[3]*Beare Essentials Ltd., Thurston, Bury St. Edmunds, United Kingdom*

[4]*Ordnance Survey, Southampton, United Kingdom*

CONTENTS

ABSTRACT This chapter will introduce how quality of geoinformation can be managed when the production environment is no longer inside one organization (e.g., collection of data is contracted out) or data are compiled from various sources, as in the case of spatial data infrastructures (SDIs). The bases for quality management of reference geoinformation are discussed using four viewpoints: data, process, organization, and user-centric. These viewpoints can be met using ISO 19157 and 19158 standards together with the European Spatial Data Infrastructure with a Best Practice Network (ESDIN)–developed quality model and Data Quality Services Framework (DQSF). Two different services are identified: a Data Producer Quality Web Service and a Data User Quality Web Service.

We discuss how these principles and services are implemented now within EuroGeographics and Ordnance Survey of Great Britain. Further development is currently implemented within the European Location Framework (ELF) project, which is providing a single source of reference geoinformation for Europe during 2013–2016.

6.1 Introduction

6.1.1 Consequences of Supply Chain Change to Data Quality

In previous times, determining spatial data-set quality, fitness for purpose, or even ease of use for data sets had been a relatively simple exercise. The data sets themselves were simple (e.g., a single point feature with one or two associated attributes). These data sets appeared even simpler, as those that had created them were very often the only organizations consuming the data. As a result, downstream systems were designed to cope specifically with that data set in mind; the data set's limitations were well understood and accounted for by most, if not all, the users within the organization.

This simple picture has of course become far more complex than many could have envisaged. The world is creating and consuming information at an alarming rate. The data have become far more complex (e.g., multifeature object-based data sets with multiple attributes with complex intrarelationships). These new data are being created (and even in some cases maintained) by multiorganization supply chains. The data are being consumed by multiple organizations for at least as many different uses. Determining quality in this complex environment requires more than planning by one or two organizations. Ensuring the most value is extracted from the data, if all are to succeed, requires a more holistic approach to quality.

6.1.2 Changing Use of Reference Geoinformation

The Infrastructure for Spatial Information in the European Community (INSPIRE) directive of the European Union[1] establishes the basis for sharing and delivering the geospatial data for environmental purposes. Annexes I and II of the directive define the reference geoinformation part, which is important for the thematic data reference (Annex III of the directive).

Reference geoinformation can be defined as the series of data sets that everyone involved with geographic information uses to reference their own data to as part of their work. They provide a common link between applications and thereby provide a mechanism for sharing knowledge and information among people.[2–4]

In Chapter 5, we noted how production of reference geoinformation is changing. Supply chains (a supply chain being a collection of processes, some within the same organization and some outside, organized together to produce one or more products) and reference geoinformation are becoming more complex. For example, in the INSPIRE context, one reference theme might contain data from several authorities.

Previously, reference geoinformation has mostly been used as a backdrop map, and then other information has been overlaid onto it. With the introduction of the linked data[5] concept and, for example, the need for connecting more attribute information to spatial data, it is important to manage change and ensure that the latest data are used. Reference geoinformation will come available through platforms, where users do not have to manage data sets, but can start integrating data as a service (DaaS)[6] for their applications. This requires a change in how data quality will be managed. One of the advantages of DaaS is that data quality can be managed uniformly, as there is a single point for updates.

6.2 Bases for Quality Management of Reference Geoinformation

Quality management of reference geoinformation must address the following:

- Provision of a cost- and time-effective and standardized framework to measure and improve quality
- Meeting of changing users' needs
- Increase of users' trust and how to create confidence in the usage of available data to make informed decisions

We introduce here three aspects for quality management of reference geoinformation that will have to be taken into account. These aspects are based on Jakobsson and Tsoulos,[7] Jakobsson,[8] and Jakobsson et al.[9]

6.2.1 Data-Centric Approach

Geoinformation quality standards ISO 19113, 19114, and 19138 have been replaced by ISO 19157: Geographic Information—Data Quality.[10] This defines the previously established data quality elements—completeness, logical consistency, positional accuracy, thematic accuracy, and temporal quality—and also introduces an additional element: usability. Basically, this enables introduction of new measures that could meet changing user requirements.

Metadata standards (like ISO 19115[11]) are often considered quality standards as well because they contain information that can be used for expressing and determining some aspects of quality. Elements like usage, lineage, and date of last revision are good examples.

These standards build an important basis for the quality management of reference geoinformation and can be considered a data-centric part of it. However, as pointed out by Devillers et al.,[12] metadata approaches have not really been as successful because of the complexity of data quality.

6.2.2 Process- and Organization-Centric Approach

ISO 9000[13] is a widely used quality management standard series that concentrates on the process- and organization-centric part of the quality management. The recently accepted technical specification ISO 19158: Geographic Information—Quality Assurance of Data Supply[14] offers a framework in which a modern supply chain can understand the quality requirement of the data being produced (or maintained).

In addition to the data-centric part, which is based on ISO 19157, the ISO 19158 technical specification considers other aspects of quality that would impact upon a supply chain, such as the schedule, volume, and cost of delivery. This approach can then be used to provide assurances that the entire supply chain is capable of producing the quality required in those terms.

The framework allows the supply chain to be broken down into its constituent processes and with particular consideration for human interaction in the data production or maintenance processes.

Using ISO 19158 to gain assurance in any given supply chain, one must first understand the user requirements. The relevant elements of quality must be identified, followed by the assignment of measures and acceptable quality levels. These must relate directly to the real customer requirement, or at least the perceived customer requirement. Once this requirement is identified, it may be used by management elements within the supply chain to identify both the required outputs and the expected inputs of individual processes. It enables the user of the specification to understand the (risk of) propagation of data error (through poor quality data being passed on to the next process), as well as the impact of poor scheduling and data volume management.

This level of understanding is achieved through a process that reviews, tests, and ensures each element of the production or maintenance processes. For any given process that impacts upon the data, there are up to three levels of assurance (basic, operational, and full). Not all levels are mandated; the levels may be considered similar to risk mitigation, and so with higher levels achieved comes greater assurance, but at a higher implementation cost.

The successful implementation of ISO 19158 is dependent on the relationship between customer and supplier, as well as the customer's understanding of the processes undertaken by the supplier. This becomes even more critical with more complex data. If it is not possible for the customer to understand and monitor the process, then a technical agent should be used to act on the customer's behalf. Note that the customer may or may not be internal to the supplier organization.

The basic level of quality assurance ensures that a process appears to be capable of creating or maintaining a product to the right quality. As this is predominantly a paper exercise, it provides the lowest level of assurance. The supplier provides appropriate evidence to the customer (or the customer's agent), which will identify its suitability for the production or maintenance of the data set. For example, documentation to be reviewed might include proposed management structure, quality plans, change control plans, training plans, tool specifications, and high-level process maps. The detail required at this level and others should be proportionate to the quality risk posed. The outcome of this assurance activity can provide information for the next level. For example, it may identify areas where quality control is critical or where there are likely data flow restrictions (bottlenecks) in the proposed production process.

The operational level of assurance comes from an assessment of a working process following implementation. Rather than the previous high-level process review approach, operational assurance looks at all relevant processes in detail and breaks them down further as required. (The requirement to do so is often identified at the basic level.) At this level, data outputs are checked for conformance to the agreed quality requirement. Also at this level, individual operators' work is assessed and ensured. In this way, all staff have a responsibility for quality, not just quality control staff. The proportion of the staff that must have achieved an appropriate level of individual assurance is agreed between the supplier and customer, allowing for staff churn and training. The responsibility for training and testing normally resides with the supplier; however, the customer (or agent thereof) is required to review achievements in terms of training records and the data quality results of their individual output. Once the appropriate number of staff has achieved the agreed standard and the output of the processes confirms that data quality, volumes achieved, and schedule adherence are acceptable, the supplier can be said to have achieved operational assurance. The testing of data (both product and individual output) may be reduced at this point as the risk diminishes.

The final level of assurance ensures that the supplier is capable of maintaining the quality achieved at the operational level over a period of time. This period will be agreed between supplier and customer. Data quality result trends will be analyzed and reviewed by management of the supplier and customer for the life of the process, with the aim of continually improving the supply chain.

6.2.3 User-Centric Approach

Usability is defined in ISO 9241[15] as "the effectiveness, efficiency, and satisfaction with which specified users achieve specified goals in particular environments." It has its roots in engineering, especially software development (e.g., Nielsen[16]). Usability has been studied in connection with geoinformation (e.g., Wachowicz and Hunter[17]).

In the previous chapters, usability is a key method for identifying the needed measures in both ISO 19157 and ISO 19158. However, in many cases, these measures, while important for evaluation of quality, are not mentioned in user interviews. Users sometimes prefer verbal results or statements from other users regarding how good the data are. From the user point of view, trust is the key factor in selection of the data or service to be used. How this trust is actually created is an interesting question. Different methods of creating trust include, for example, certification (as in the case of ISO 9000 certification), accreditation (e.g., ISO 17025 for laboratory testing[18]), and now ISO 19158, which introduces assurance levels. Other examples include quality labeling, for example, GEO label for GEOSS,[19] and quality visualization.

Using authoritative sources also creates trust for users, and this is especially important for reference geoinformation because most of the reference data are produced by public agencies. An authoritative source is "a managed repository of valid or trusted data that is recognized by an appropriate set of governance entities and supports the governance entity's business environment."[20] The challenge here is that even if the source is considered authoritative, it may lose users' trust if it does not deliver good quality.

6.3 Quality Management of Reference Geoinformation in a Multi-Data-Set SDI Environment

The above principles have not yet been implemented into a multi-data-set SDI environment. In the European Spatial Data Infrastructure with a Best Practice Network (ESDIN) project, ISO 19157 principles were utilized in the Data Quality Services Framework (DQSF) (Figure 6.1). This framework contains services for data supplier and data user. Further development of this DQSF is currently implemented in the ELF project discussed later in Section 6.5.1.

ISO 19158 has not yet been tested in an SDI context. Full interoperability relies on all parties having the same perspective on data quality; however, the implementation of ISO 19158 within an SDI would ensure that data purpose and quality are at least understood by all relevant parties. The proliferation of discovery quality metadata (aggregated from the data quality results identified in the assurance process) would provide this opportunity.

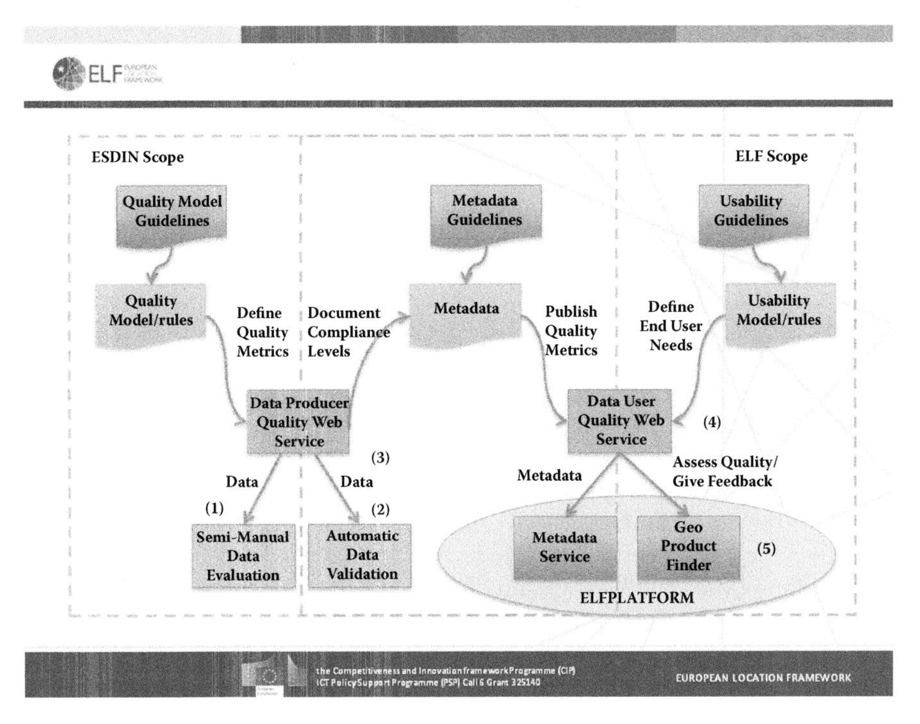

FIGURE 6.1
Data Quality Web Service Framework. (Modified from ESDIN, Public final project report, ESDIN D1.10.3, 2011, http://www.esdin.eu.)

6.3.1 ESDIN Work

The recently finished ESDIN project[21] made some fundamental findings in the quality management of reference geoinformation. The project's main focus was to study how to implement the INSPIRE directive for reference geoinformation, and its central findings on quality were as follows:

- An integrated model of quality and quality measures can be created for reference geoinformation (quality model).
- Quality validation can be automated as a rule (DQSF).

When these results are put into practice, they will fundamentally change the quality management of reference geoinformation.

6.3.2 Approach

The ESDIN approach is illustrated in Figure 6.1 (see ESDIN scope part).[22,23] It uses ISO 19157 as a framework for evaluating data quality using quality measures. It includes parts that can be checked automatically, like conformance rules, and parts that require manual checking, like completeness and positional accuracy. A quality model has to be defined for each data set. This will set the quality requirements using the quality measures (from ISO 19157). Quality requirements should be set using users' requirement studies.

After running an evaluation using the Data Producer Quality Web Service, results may be reported in metadata. The most common case would be data set-level metadata for the feature types and attribute types reporting conformance levels set in the quality model. These conformance levels are validated by a semimanual service, typically through sampling, but actual test results are not reported in metadata. Typically, these measures are related to completeness, positional accuracy, and thematic accuracy. For logical consistency and temporal accuracy, actual test results may be provided, as the whole data set may be tested automatically.

In the ESDIN project, a need for the Data User Quality Web Service was recognized. This builds on setting the usability model based on user requirements, and then the Data User Quality Web Service will give advice on whether the data meet the user requirements. In the ELF project, an implementation of the Data User Quality Web Service will be developed.

6.4 Use Cases

6.4.1 EuroGeographics

EuroGeographics[24] is a not-for-profit organization representing 56 national mapping, land registry, and cadastral authorities (NMCAs) in 45 European countries. It has a long experience in building harmonized data sets based on its members' data.

6.4.1.1 ERM Quality Control

EuroRegionalMap (ERM) is a topographic data set covering 35 countries in Europe and based on NMCA data at the European regional level of detail (1:250,000). While technical interoperability can be ensured by the use of a common data model (ERM data model), it is more challenging to provide comparable and harmonized data content. As shown in Pammer et al.,[25] the national production workflows vary due to national constraints, resources, and the availability and accessibility of suitable data sources. Mainly, national specifics cause deviations from the ERM specification with respect to the selection criteria, level of generalization, and quality.[26]

The ERM data specification provides a description of the content, accuracy, and data. The quality requirements are indicated as general requirements (absolute horizontal accuracy, data density level, and selection criteria;

dimensions: geometric resolution) and portrayal and quality criteria by feature type.

In order to ensure good quality of the resulting ERM product, the ERM validation specification details the validation procedure that should be carried out throughout the production process:

- The national producers are responsible for the validation of their national contribution using, whenever possible, the validation tools implemented for the final data validation and assessment phase. It is the responsibility of the data producer to ensure completeness of data collection.
- ERM regional coordinators perform the final validation and quality assurance (QA) of the national data components for final acceptance.
- A final validation and QA are carried out after the data assembly phase on the full European data set by the production management team.

The validation procedures consist of a series of checks to identify errors in the data's geometrical and topological structure, as well as feature and attribute compliance with the current ERM specifications and consistency with the data collection. The current process is mainly focused on supporting the production management. The validation results have been returned to each producer with recommendations on how the national contribution can be improved for the next release.

6.4.1.2 Identifying Risks in the Production Processes

Key to both ISO 9000 and ISO 19158 is the need for data managers to be able to demonstrate they know what data processes exist in the supply chain, when they are applied, what they are seeking to achieve, and the level of risk associated with achieving the desired outcome based on how they are applied. Thereby data managers need to determine the type and level of quality controls required at each stage in the supply chain. In the ESDIN project, a high-level process of collating national data into pan-European data was identified, highlighting where the data quality issues had the greatest concern or impact on the supply chain (Figure 6.2).

The point of impact on the operation is felt most by the regional coordinators, when evaluating the quality of the received data as part of the data assembly process. Issues often arise with respect to nonconformance to the data specifications (DSs) and to edge matching (EM) errors across national boundaries. This results in the need to raise issue reports with the NMCAs for the data to be corrected. This correction cycle can be repetitive and inefficient and causes significant delay to the production cycle of the EuroGeographics

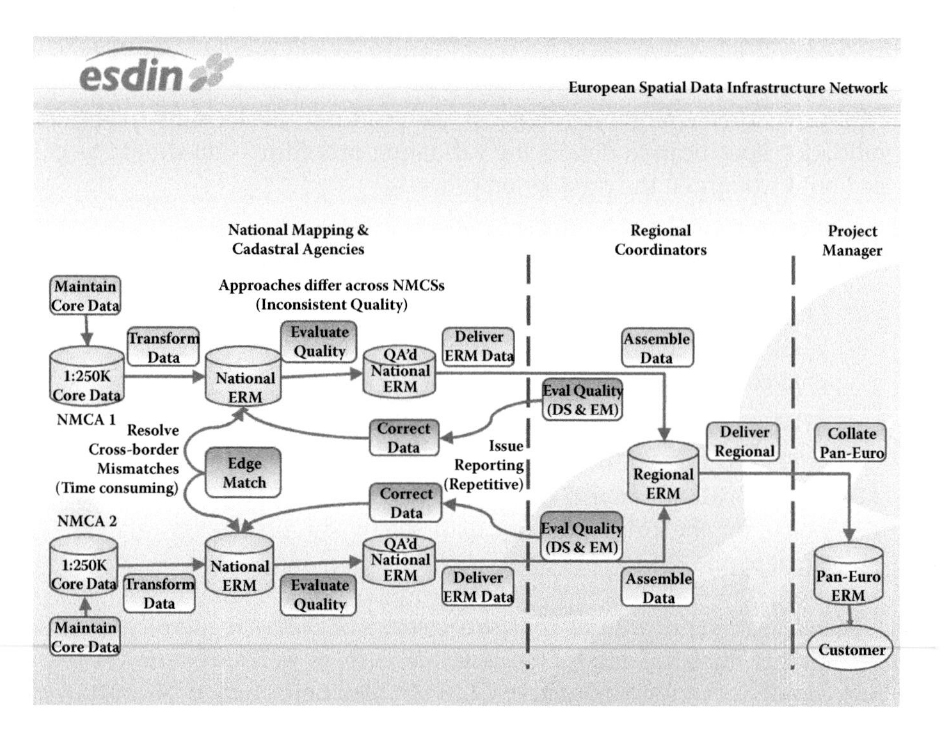

FIGURE 6.2
ERM production process and identified risks.

products. Indeed, overseeing this correction process for data assembly is the primary reason for needing regional coordinators in the first place.

The solution proposed by the ESDIN project was to introduce an edge matching service based on using a common International boundary data set and connecting features. The other significant improvement suggested was to introduce a common data quality service that can be used at both the national and the European level (Figure 6.3).

6.4.1.3 ERM Quality Assurance

Since the successful completion of the ESDIN project, the ERM production management team has started to apply the guidance of the ESDIN quality model for the ERM data set.

With the assistance of 1Spatial, the collation of national data contributions has been enhanced through the introduction of an automated data quality evaluation process. This process has enabled full data-set evaluation for transport, hydrography, and settlement themes from all 32 national contributors.

Providing uniform assessment against a common set of (around 200) quality measures (business rules), quantitative and comparative metrics were automatically compiled for each national data set, with coverage of geometric

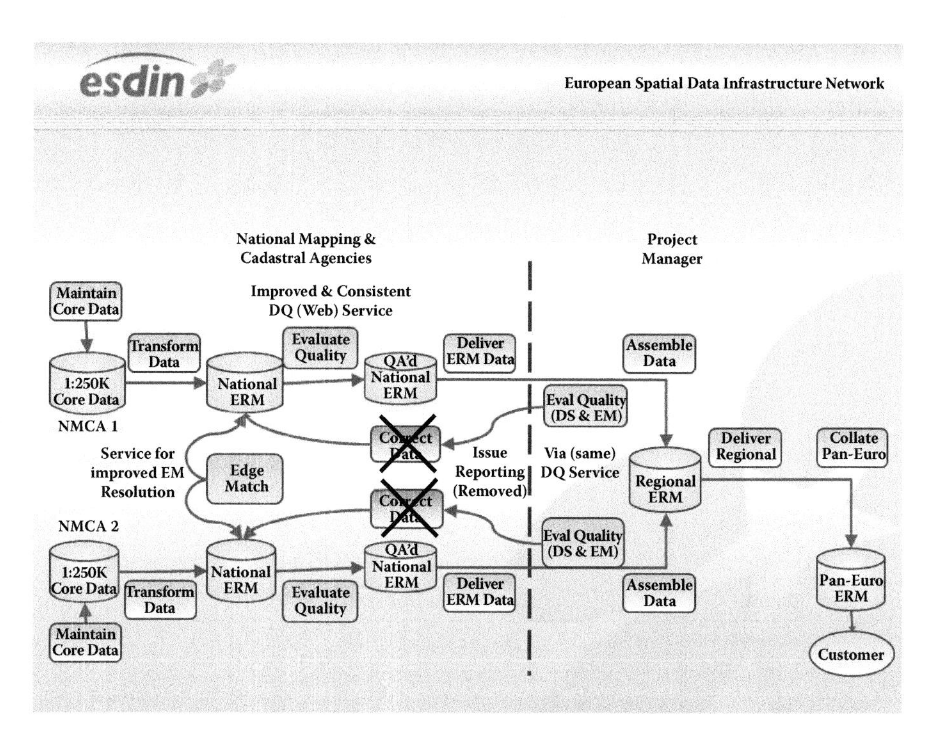

FIGURE 6.3
Improved ERM process.

resolution, domain value integrity, and topological connectivity (including cross-border consistency). These quantitative metrics provided objective viewpoints on the comparative quality of national contributions and awareness of the consistency of the data product across the whole of Europe.

This knowledge helped to increase the confidence levels that the EuroGeographics ERM production management team have in the product quality prior to distribution to customers. Additionally, detailed nonconformance reports provided the management team with the information needed to advise data providers where they could best utilize their resources to improve data quality for future product releases, thereby instilling an informed continuous improvement process.

The discussion regarding conformance levels resulted in the definition of three conformance classes and corresponding acceptance levels based on the ERM quality criteria, as shown in Table 6.1.

Further, a first ERM layer has been evaluated using the above conformance classes. This means the validation results for each ERM quality criterion have been assessed and assigned an acceptance level. Then the results have been grouped by data quality element to come up with an overall quality grading by country following the grade proposed by the ESDIN project (Table 6.2).

TABLE 6.1

Proposed Conformance Classes

	Conformance Levels		
Criteria	**Description**	**Class**	**Acceptance Level**
Shall be fulfilled	Basic requirements of ERM dataset	A	100%
		C	<100%
To be fulfilled	Basic requirements for user purpose (to be described for each item, see update plan requirements)	A	100%
		B	>90%
		C	<90%
Nice to have	General requirements of user—a goal that needs to be achieved (i.e., the attribute completeness rate for mandatory)	A	100%
		B	>75%
		C	<75%

TABLE 6.2

Grading Data (ESDIN 2011)

Grade	Data Quality Description
Excellent	Only class A for all quality measures
Very good	A majority of A's, but also some B's
Good	A majority of B's, some A's, no C's
Adequate	Only a very few C's, the other B's and better
Marginal	A majority of C's but also some B's
Not good	No measure reached the class B (i.e., all measures on class C)

Source: ESDIN, Public final project report, ESDIN D1.10.3, 2011, http://www.esdin.eu.

The discussion of the results in the ERM production management team showed that the aggregation of data quality results poses some issues regarding the following:

- Aggregation where measurements are at different scales and units
- Aggregation for inhomogeneous data
- Reporting details

The pilot implementation of the ESDIN quality model proved the applicability of the proposed ESDIN data quality model. However, a good understanding of the International Organization for Standardization (ISO) standards on data quality is required. Furthermore, the objective of the quality reporting, that is, report to producers, management, or users, needs to be clearly defined.

Also, application of ISO 19158 has been started for the ERM producers. First, a basic-level assurance has been achieved by the new NMCAs joining the EuroGeographics' production program.

6.4.2 Ordnance Survey

Ordnance Survey implemented its approach to quality assurance (a forerunner to ISO 19158) following the experiences it had gained from letting contracts over the years.

Prior to implementation, Ordnance Survey had let contracts to maintain its large-scale database with limited input to the specific processes, tools, and individuals that would be updating data on their behalf. As data sets became more complex, Ordnance Survey and its suppliers started to experience data quality issues. Through the application of the approach outlined above, Ordnance Survey was able to better support its suppliers, and in return, it received the quality of data that they it, to the appropriate volume and schedule. Realizing the benefit of this approach, Ordnance Survey then applied it to all internal production and maintenance processes.

The approach has been successful in identifying data quality issues early in the process development cycle, providing opportunity for Ordnance Survey to work with its suppliers to resolve those issues before they become unmanageable. As data sets have become even more complex, there is greater opportunity for this approach to add value. With this approach the customer, supplier, and individual operator have a good understanding of the data quality that is required and the quality that is being produced. As the relationship between the two is continually monitored, it may be managed proactively and effectively.

There are challenges to be overcome. For example, many consider that there is no customer value in quality metadata: to the end user, the value lies in the data themselves. As data sets become more complex, more assurance is potentially loading processes and individuals with essential but non-value-adding costs and at the same time adding precious time to the process. As a result of this, there is a tipping point at which individuals will become disenfranchised with the production and maintenance process, which in turn will have a negative effect on data quality. This challenge may be mitigated with the investment in automated testing, as discussed earlier.

6.5 Further Development

6.5.1 European Location Framework Project

During the next 3 years, the European Location Framework (ELF) project will deliver the first implementation of the European Location Framework[27]—a technical infrastructure that harmonizes national reference data to deliver authoritative, up-to-date, interoperable, cross-border geospatial reference data for use by the European public and private sectors in a way that is easy to use by application developers and even end users.

The project will provide a critical mass of content and coverage as 15 member states' national ELF and INSPIRE data will be made available from a single source (ELF platform), connecting to a number of applications, the European Commission INSPIRE geoportal, the commission internal portal run by Eurostat, and ArcGIS Online, a commercial cloud geographic information system (GIS) platform.

The ELF platform will be implemented using an open-source development made originally for the Finnish SDI, Oskari. Covering the full range of INSPIRE Annex I, II, and III themes, these data sets will provide full national coverage of the rich content available from national and regional spatial data infrastructures.

In the ELF project, quality evaluation based on ESDIN results will be operationalized using cloud-based commercial services. The goal is also to introduce a standard way in which quality models can be expressed as rules, which enables using these in multiple software environments. The project has decided to use RuleSpeak[28] for expressing the rules in a standardized, technology-agnostic way. Figure 6.2 represents the ELF data quality approach. It is based on implementing the Data Quality User Web Service presented in Figure 6.1.

Compared to the ESDIN approach (see Figure 6.4), ELF separates national requirements and processes from European ones. The ESDIN principles may

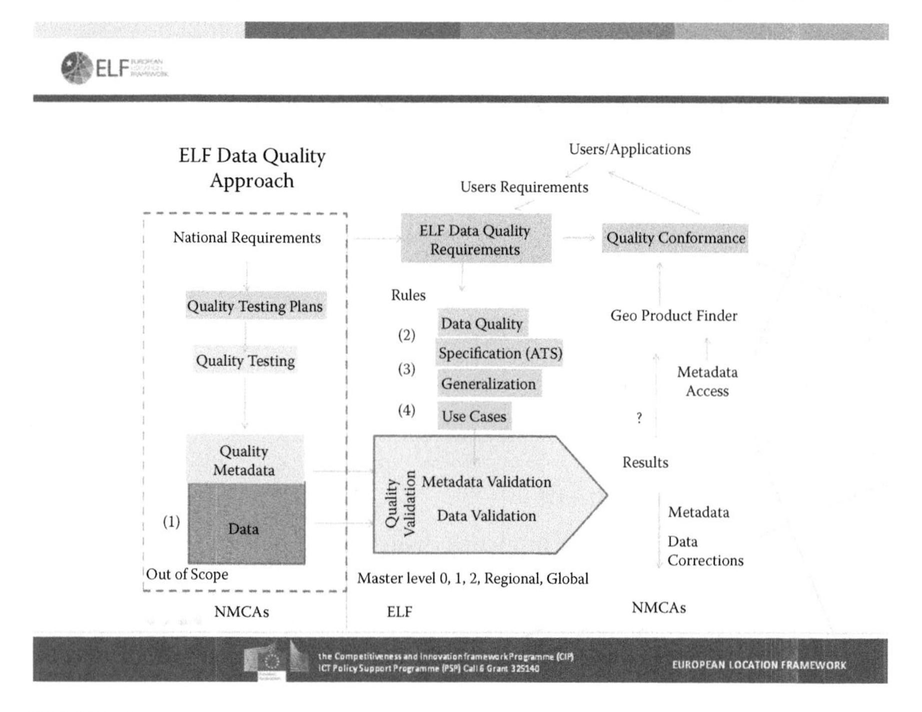

FIGURE 6.4
ELF data quality approach.

be used for both processes, but there are different requirements. Therefore, within ELF the common requirements are defined, which then are going to be validated. The national quality metadata are taken into account in the validation process. The rules used for the validation can be classified to different categories, for example, meeting generalization or other use cases. The Geo Product Finder service will then be used for reporting quality results and getting user feedback.

ELF, with its partnership approach to the customer–supplier relationship and the implementation of ISO 19158, offers opportunities in organizational interoperability, encouraging organization and process alignment. This alignment can lead to opportunities at the technical level (particularly around the resolution of data quality issues). It supports the requirement for quality metadata for discovery purposes, and given the findings of the ESDIN project, it also supports the approach and provides a framework for repetition of the quality evaluation process for all process steps in SDIs.

6.6 Summary

In this chapter, we have presented an approach for management of data quality within reference geoinformation. This is based on the ESDIN Data Quality Service Framework, which specifies two data quality validation services: a Data Producer Quality Web Service, which may be utilized by the data producer or owner, and a Data User Quality Web Service, which may be used by the data user or manager (e.g., in a multi-SDI situation). Automation of data quality validation processes has been one of the main drivers of this work. EuroGeographics European data sets have been used for testing the data quality validation process. Based on ESDIN recommendations, data quality aggregation has been applied for presenting the validation results. Implementation of these services is currently progressing in the ELF project.

We also introduced organization and process viewpoints to the management of quality. ISO 19158 introduces quality assurance levels that can be applied to data production processes and organizations. Ordnance Survey has successfully implemented this in their data supply processes, and first implementations have been done within EuroGeographics.

The main drivers for introducing better-quality management to reference geoinformation production are partly based on government policies like e-government, legislation (e.g., INSPIRE directive), and cost-effectiveness, and partly on users' demands. We believe that introduction of the ESDIN Data Producer Quality Web Service and the ESDIN data quality model in the national services and the ELF Data User Quality Web Service for the European and cross-border data within the ELF project will decrease production cost and time and increase trust. This will enable faster and more frequent release of reference geodata at the national, regional, and global levels. Further, introduction of ISO 19158 at the national but also at the international level will increase users' trust to reference geoinformation.

References

1. European Union. 2007. Directive 2007/2/EC of the European Parliament and of the Council of 14 March 2007 establishing an Infrastructure for Spatial Information in the European Community (INSPIRE). http://eur-lex.europa.eu/LexUriServ/LexUriServ.do?uri=CELEX:32007L0002:EN:NOT (accessed April 27, 2013).
2. Federal Geographic Data Committee (FGDC). 2005. *Framework Introduction and Guide* [handbook]. Digital version. http://www.fgdc.gov/framework/frameworkintroguide/ (accessed April 27, 2013).
3. Rase, D., Björnsson A., Probert, M., and Haupt, M.-F. (eds.). 2002. Reference data and metadata position paper. Eurostat. http://inspire.jrc.ec.europa.eu/ (accessed January 26, 2012).
4. Global Spatial Data Infrastructure Association (GSDI). 2009. *The SDI Cookbook.* http://www.gsdidocs.org/GSDIWiki/index.php/Main_Page (accessed April 27, 2013).
5. Berners-Lee, T. 2009. Linked data. http://www.w3.org/DesignIssues/LinkedData.html (accessed April 27, 2013).
6. Wikipedia. 2014. (accessed May 2, 2014).
7. Jakobsson, A., and Tsoulos, L. 2007. The role of quality in spatial data infrastructures. In *Proceedings of the 23rd International Cartographic Conference,* Moscow, Russia, CD-ROM.
8. Jakobsson, A. 2006. On the future of topographic base information management in Finland and Europe. Doctoral dissertation, University of Technology, Helsinki. http://lib.tkk.fi/Diss/2006/isbn9512282062/ (accessed April 27, 2013).
9. Jakobsson, A., Mäkelä, J., Henriksson, R., Tsoulos, L., Beare, M., Marttinen, J., and Lesage, N. 2009. Quality beyond metadata—Implementing quality in spatial data infrastructures. In *Proceedings of the 24th International Cartographic Conference,* Santiago de Chile, Chile, CD-ROM.
10. International Organization for Standardization. 2013. ISO 19157: Geographic information—Data quality. ISO, Geneva.
11. International Organization for Standardization. 2014. ISO 19115-1: Geographic information—Metadata—Part 1: Fundamentals. ISO, Geneva.
12. Devillers, R., Stein, A., Bédard, Y., Chrisman, N., Fisher, P., and Shi, W. 2010. 30 years of research on spatial data quality: Achievements, failures and opportunities. *Transactions in GIS,* 14(4): 387–400.
13. International Organization for Standardization. 2005. ISO 9000:2005: Quality management systems—Fundamentals and vocabulary. ISO, Geneva.
14. International Organization for Standardization. 2012. ISO/TS 19158: Geographic information—Quality assurance of data supply. ISO, Geneva.

15. International Organization for Standardization. 1997. ISO 9241-1:1997: Ergonomic requirements for office work with visual display terminals (VDTs)—Part 1: General introduction. ISO, Geneva.
16. Nielsen, J. 1993. *Usability Engineering*. Academic Press, Boston.
17. Wachowicz, M., and Hunter, G. 2003. Spatial data usability, preface. *Data Science Journal* (Spatial Data Usability Special Section), 2. https://www.jstage.jst.go.jp/article/dsj/2/0/2_0_75/_article (accessed April 27, 2013).
18. International Organization for Standardization. 2005. ISO 17025:2005: General requirements for the competence of testing and calibration laboratories. ISO, Geneva.
19. Geoviqua. http://www.geoviqua.org/GeoLabel.htm (accessed May 2, 2014).
20. Westman, R. 2009. What constitutes an authoritative source? In A. Durrant et al. (eds.), *Proceedings of the Privilege Management Workshop*, September 1–3. http://csrc.nist.gov/news_events/privilege-management-workshop/presentations/Roger_Westman.pdf (accessed April 27, 2013).
21. ESDIN. 2011. Public final project report. ESDIN D1.10.3.http://www.esdin.eu (accessed April 27, 2013).
22 Beare, M., Henriksson, R., Jakobsson, A., Marttinen, J., Onstein, E., Tsoulos, L., Williams, F., Mäkelä, J., De Meulenaer, L., Persson, I., and Kavadas, I. 2010. Quality final report—Part A. ESDIN D8.4.http://www.esdin.eu/sites/esdin.eu/files/D8-4_ESDIN_Quality_Final_Report.pdf (accessed April 27, 2013).
23. Jakobsson, A., Beare, M., Marttinen, J., Onstein, E., Tsoulos L., and Williams, F. 2011. A cohesive approach towards quality assessment of spatial data and its automation. In *Proceedings of the 25th International Cartographic Conference*, Paris, CD-ROM.
24. EuroGeographics. http://www.eurogeographics.org (accessed April 27, 2013).
25. Pammer, A. et al. 2009. EuroRegionalMap: How to succeed in overcoming national borders. In G. Gartner and F. Ortag (eds.), *Lecture Notes in Geoinformation and Cartography: Cartography in Central and Eastern Europe*. 1st ed. Springer, Berlin, pp. 19–40.
26. Hopfstock et al. 2012. EuroRegionalMap: A joint production effort in creating a European topographic reference dataset. In M. Jobst (ed.), *Service Oriented Mapping 2012*. JOBST Media Präsentation Management, Verlag, Wien, pp. 151–162.
27. Jakobsson, A. 2012. Introducing a new paradigm for provision of European reference geo-information: Case study of the European Location Framework concept. In M. Jobst (ed.), *Service Oriented Mapping 2012*. JOBST Media Präsentation Management, Verlag, Wien, pp. 51–62.
28. Ross, R.G. 2009. Basic RuleSpeak guidelines: Do's and don'ts in expressing natural-language business rules in English. Version 2.2. Business Rule Solutions, Houston, TX, USA.

7

A New Approach of Imprecision Management in Qualitative Data Warehouse

F. Amanzougarene,[1,2] M. Chachoua,[1] and K. Zeitouni[2]

[1]*EIVP: École des ingénieurs de la ville de Paris, Paris, France*

[2]*PRISM, Université de Versailles–SQ, Versailles, France*

CONTENTS

ABSTRACT Data warehouse means a decision support database allowing integration, organization, historization, and management of data from heterogeneous sources, with the aim of exploiting them for decision making. Data warehouses are essentially based on a multidimensional model. This model organizes data into facts (subjects of analysis) and dimensions (axes of analysis). In classical data warehouses, facts are composed of numerical measures and dimensions that characterize them. Dimensions are organized into hierarchical levels of detail. Based on the navigation and aggregation mechanisms offered by online analytical processing (OLAP) tools, facts can be analyzed according to the desired level of detail. In real-world applications, facts are not always numerical and can be of a qualitative nature. In addition, sometimes a human expert or learned model such as a decision tree provides a qualitative evaluation of the phenomenon based on its different parameters, that is, dimensions. Conventional data warehouses are thus not adapted to qualitative reasoning and do not have the ability to deal with qualitative data. In previous work, we have proposed an original approach for qualitative data warehouse modelling that permits integrating qualitative measures. Based on computing with words methodology, we have extended the classical multidimensional data model to allow the aggregation and analysis of qualitative data in the OLAP environment. In this chapter, we focus our study on the representation and management of imprecision in the annoyance analysis process.

7.1 Introduction

Data warehouses and online analytical processing (OLAP) constitute the main elements of decision support systems. A data warehouse means a decision support database allowing integration, organization, historization, and management of data from heterogeneous sources, with the aim of exploiting them for decision making (Kimball and Ross, 2002; Inmon, 2005). OLAP refers to the technology that allows users to efficiently retrieve the information stored in a data warehouse. To conceptualize data in a data warehouse, the multidimensional model is used. This model organizes data into facts (subjects of analysis) and dimensions (perspectives of analysis). A fact is composed of numerical measures and dimensions that characterize it. A dimension is organized into hierarchical levels of detail. Based on the navigation and aggregation mechanisms offered by OLAP tools, facts can be analyzed according to the desired level of detail. In some real-world applications, the subject of analysis may be subjective, and consequently, its measures are provided in a qualitative fashion. In addition, sometimes a human expert or a prediction model such as a decision tree can be used to provide a qualitative evaluation of some phenomenon based on its different parameters. This arises in many applications, such as customer satisfaction, process control, consumer products, and annoyance evaluation. Conventional data warehouses are thus not adapted to human reasoning and do have the ability to deal with qualitative data. In previous work, we have presented an original work that aims to make it possible to handle raw qualitative measures and provide a more flexible method for multidimensional analysis over that type of data. Based on computing with words methodology, we have introduced qualitative measures and aggregates as an extension of the multidimensional data model of a data warehouse. Using these measures and aggregates, OLAP queries allow the decision maker to manipulate data in a qualitative fashion using linguistic terms. In this chapter, we extend this model to deal with both qualitative and quantitative measures, which leads to handling imprecise data in a data warehouse. Compared to the state of the art, there exist several research works addressing aggregation over imprecise and uncertain data (e.g., Laurent, 2001; Molina et al., 2006; Burdick et al., 2007; Delgado et al., 2007). Our study focuses on the fuzzy fusion of qualitative and quantitative measures in the context of data warehouses. To illustrate the problem and our solution, we consider throughout this chapter the annoyance of urban building sites.

This chapter is structured as follows. In Section 7.2, we present our work motivation and the use case related to urban building site annoyance evaluation and analysis. In Section 7.3, we propose a data model allowing the combination of qualitative and quantitative measures in the context of imprecise multidimensional databases. In Section 7.4, we present the experimental framework designed for the annoyance analysis. Finally, in Section 7.5, we conclude and present some perspectives.

7.2 Motivation and Use Case: Urban Building Site Annoyance

Although indispensable for the development and renovation of cities, urban building sites are often a source of various kinds of nuisance. These nuisances do not have negligible impacts on the quality of life of urban citizens. This issue is crucial and becomes more complex in cities with high population density. We make the observation that in human reasoning, the annoyance is evaluated subjectively and qualitatively by using an ordinal scale of linguistic degrees. Therefore, for a perfect match with the human expert reasoning, we propose in this chapter a qualitative model of annoyance evaluation. In our previous studies (Amanzougarene et al., 2012), we have presented a quantitative model that allows evaluating urban people annoyance due to the noise. By comparison, in the present work, we generalize our previous model by privileging a qualitative data handling of annoyance. We also extend our previous model of annoyance evaluation to other types of nuisance than noise, which strengthens the interest of multidimensional analysis. Indeed, an urban building site is generally likely to cause many nuisances.

7.2.1 Qualitative Representation of Annoyance

7.2.1.1 Notion of Annoyance

As several studies show it, the annoyance is an unpleasant sensation experienced by an individual facing deterioration in the quality of her environment

TABLE 7.1

Main Factors of Annoyance

Factors Related to Individual	Factors Related to Nuisance	Factors Related to Environment
Age	Nuisance type	Space
Health condition	Nuisance intensity	Time
Gender	Exposure duration	
Socioprofessional category	Frequency	
Acceptability		
Awareness		

(Guski and Felscher-suhr, 1999; Nordin and Lidén, 2006; Moser and Robin, 2006; Robin et al., 2007). The annoyance may be caused by various nuisances (noise, odor, vibration, traffic congestion, air pollution, etc.). According to various factors (intensity, moment, type, etc.), a nuisance is likely to cause a more or less important annoyance to individuals. Note that the level of annoyance caused by one or more nuisances can be different from one individual to another, depending on various factors (sensibility, age, acceptability, etc.). This means that a phenomenon that is not at all annoying for an individual can be extremely annoying for another individual. That reflects the subjective character of the annoyance notion. Thus, for the rest of our study, we propose the following definition for the annoyance notion:

> In a spatiotemporal environment, annoyance is the subjective relationship between an individual and a harmful phenomenon.

In other words, an individual can only be annoyed in the presence of one or more harmful phenomena for this individual. Thus, a human expert can subjectively evaluate the degree of annoyance, according to the various factors (Amanzougarene et al., 2012). The most relevant factors can be classified in three categories: (1) factors related to the individual, (2) factors related to nuisance, and (3) factors related to the environment. Table 7.1 shows these three categories with the main factors.

These three categories of factors will be used to evaluate the annoyance and determine the scenario of building site that produce the minimum of annoyance.

7.2.1.2 Dimensions of Annoyance

In practice, the choice of factors to be considered for the annoyance evaluation depends on the human experts' appreciation. In our case study, the experts have retained some factors related to the individual, nuisance, and environment. The latter is actually a combination of space and time dimensions. This leads to a multidimensional representation described by Figure 7.1 and including the following dimensions:

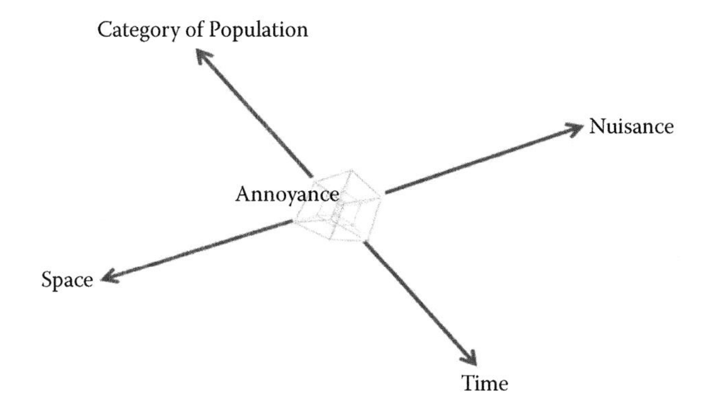

FIGURE 7.1
Multidimensional representation of annoyance.

1. Category of population grouping the factors related to the individual
2. Nuisance grouping the factors related to the nuisance
3. Space
4. Time

Note that the choice of the dimensions is application dependent and could add or ignore some factors, such as the building type or gender. Our model adapts to other schemas as well.

Categories of population: In our case study, the categories of population exposed to nuisances are represented by a typical individual. These are determined by a combination of factors already presented in Table 7.1. For instance, a category of population could be "healthy senior manager," which means implicitly an adult individual of male gender whose socioprofessional category is manager, in good health condition. Another category of the population could be "housewife mother," meaning an unemployed adult of female gender, who has one or more children. A third one could be "child's breathing problems," meaning a young individual who is not in good health condition.

Nuisances: The different categories of population concerned with the carrying out of an urban building site are exposed to nuisances that can be classified into three categories: (1) nuisances related to the living environment, characterizing unwanted changes in habits of the population impacted; (2) nuisances related to the landscape insertion of urban building sites, describing changes in the visual environment; and (3) sensorial nuisances, such as noise, dust, odors, and vibration.

Time: People are not annoyed in the same way according to the moments of the day and the periods of the year. For example, given residential area, a loud noise can be accepted during the day, but not at all at night. In our case study, we define a hierarchy of time. This hierarchy consists of dividing the year into two periods: (1) rainy period and (2) nonrainy period. The weekdays are divided into three moments: morning, evening, and night.

Space: The annoyance of an individual may vary depending on his or her distance to the source of nuisance. Indeed, nuisances generated by urban building sites are not present in a uniform manner inside the influence area. It is thus important to decompose this area into several subareas, for example, the immediate vicinity, the influence area, and the boundary of the influence area. For this dimension, we will use a priori geographical zoning.

7.2.2 Annoyance Evaluation

In human reasoning, the subjective evaluation of annoyance is done qualitatively by using a finite scale of linguistic degrees, such as low, high, and so on. Generally, the human subject uses ordered scales with five or seven linguistic degrees (Yager, 2007). In our case study, the evaluation process is as follows (for the sake of space, we will briefly describe it):

1. Define four combinations of dimensions: category of population–time, time–nuisance, category of population–nuisance, and intensity–nuisance.
2. For each combination, a scale of 1 to 4 is used.
3. The value of a given evaluation is the product of the values corresponding to the precedent combinations.
4. Thus, the interval of evaluation is [1–256]. This interval is divided into five subintervals: [0–10], [10–30], [30–60], [60–100], and [100–256]. To these subintervals, the following linguistic terms are, respectively, associated: {insignificant, low, medium, high, extreme}.

NOTE: Hereafter, in the interests of simplifying notations, we will represent these linguistic degrees, respectively, by α_i

Example 7.1: Annoyance Evaluation

Let us consider a given location *L*1 where one has three nuisances: noise, odor, and dust. An extract of the annoyance evaluation carried out by the human experts is shown in Table 7.2.

We note that, this evaluation takes into account only the following factors: (1) socioprofessional category (SPC), (2) age, (3) type, (4) intensity, (5) time of day, and (6) period of year. In this evaluation, five levels of nuisance intensity are considered. Level 1 corresponds to the absence

TABLE 7.2

Application: Example of Annoyance Evaluation

<table>
<tr><td colspan="3" rowspan="3">Space = L_1</td><td colspan="4">Category of Population</td></tr>
<tr><td>SPC</td><td colspan="3">Inactive Resident</td></tr>
<tr><td>Age</td><td>Young</td><td>Adult</td><td>Old People</td></tr>
<tr><td>Nuisance</td><td>Type</td><td>Intensity</td><td colspan="4"></td></tr>
<tr><td rowspan="5"></td><td rowspan="5">Noise</td><td>1</td><td rowspan="5"></td><td>α_1</td><td>α_1</td><td>α_1</td></tr>
<tr><td>2</td><td>α_1</td><td>α_2</td><td>α_3</td></tr>
<tr><td>...</td><td>...</td><td></td><td></td></tr>
<tr><td rowspan="2"></td><td>α_3</td><td>α_3</td><td>α_4</td></tr>
<tr><td>α_5</td><td>α_5</td><td>α_5</td></tr>
<tr><td colspan="3" rowspan="3"></td><td>Time</td><td>Morning</td><td>Evening</td><td>Night</td></tr>
<tr><td>Period</td><td colspan="3">Nonrainy Period</td></tr>
<tr><td colspan="4">Time</td></tr>
</table>

of nuisance, which means that the degree of annoyance is α_1, that is, not at all annoyed. This table is an extract of the decision matrix carried out by the experts based on different dimensions of annoyance. This matrix will serve as a knowledge base to populate the data warehouse designed to contain data related to annoyances. This warehouse constitutes the core of our SDSS.

7.3 Toward a Qualitative Multidimensional Model for Handling Imprecise Data

In this section, first, we describe the multidimensional model of annoyance that will be used as a running example for the rest of the chapter. Then, we describe our proposed model to represent and manage imprecision in the context of the data warehouse.

7.3.1 Multidimensional Data Model of Annoyance

In our case study concerning urban building sites, the subject of analysis corresponds to the annoyance. This subject is analyzed according to the dimensions we have presented in Section 7.2.2: nuisances and categories of population, time, and space. To model data of urban building sites, we have

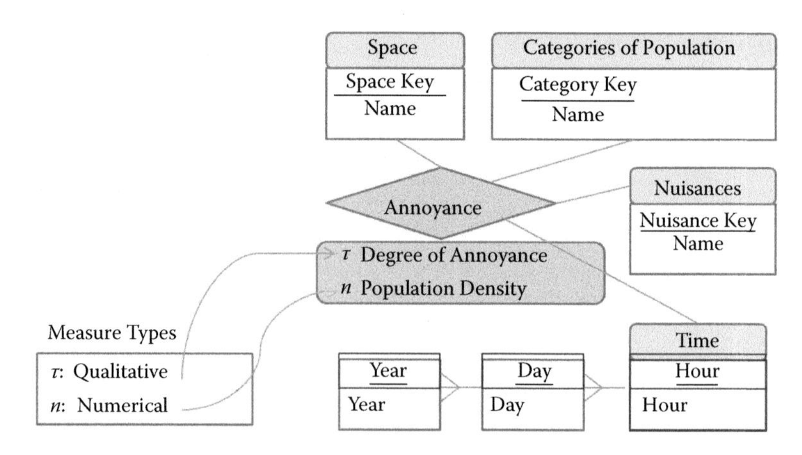

FIGURE 7.2
Multidimensional data model of annoyance.

used a star schema represented by Figure 7.2. **For the sake of simplicity, we omit the dimension attribute details in the figure.** To represent it, we use the graphical formalisms proposed by Malinovski (2008). We have defined an annoyance fact table. Thus, data from the annoyance table are analyzed according to the following dimensions: nuisances and categories of population, time, and space. Measures associated with the annoyance fact table are degree of annoyance and population density.

In the current model, we are faced with two problems:

1. The model expert used to evaluate the degree of annoyance provides a qualitative crisp value. Thus, this model does not capture the imprecision inherent to this measure.
2. The fusion of measures. Indeed, the managers of public spaces are interested in the analysis of the impact of annoyance. This measure is derived from the degree of annoyance, which is a qualitative measure, and population density, which is a numerical one. We recall that the degree of annoyance is the annoyance level of an individual type representing a given category of the population. Thus, we define the impact of annoyance as the overall level of annoyance of a given category of population, taking into account the density of this category.

For the first problem, we propose using a fuzzy set to represent the imprecision that is inherent to data. For the second problem, we propose using the fuzzy fusion approach, which is based on fuzzy inference.

7.3.2 Fuzzy Data Model

In this section, we present the proposed data model focusing on the imprecise qualitative measures. This model is based on the fuzzy theory introduced by Zadeh (1975). This choice is justified by the following:

1. Fuzzy theory offer techniques that allow us to model imprecise or vague knowledge that very few other methodologies allow.
2. It is the only way to treat in the same part the knowledge provided digitally (by instruments) and the knowledge expressed symbolically (by a human observer).

The model that we propose to represent imprecise measures is based on the concept of linguistic variable introduced in Zadeh (1975).

A linguistic variable is a 3-tuple (N, X, T_L), where one has

- N = a symbolic variable defined of X
- X = universe of discourse
- $T_L = \{A_0, A_1, \ldots, A_n\}$, a set of labels (terms linguistics)

Each label corresponds to a fuzzy set represented by a trapezoidal membership function. Let x be an imprecise value of type N. We write "x is N" and use the following general form to represent it:

$$N(x) = (\mu_{A_0}(x)/A_0, \mu_{A_1}(x)/A_1 \ldots \mu_{A_n}(x)/A_n$$

where $\mu_{A_1}(x)$ is the membership degree of to the fuzzy set A_i. For the sake of simplicity, we can write

$$N(x) = (\mu_{A_i}x)$$

7.3.2.1 Application

Representation of annoyance degree: To represent the annoyance degree measure, we propose using the following linguistic variable:

$$(DoA, R^+, T_{DoA})$$

where DoA is the name of the variable, R^+ is the universe of discourse, and T_{DoA} = {insignificant, low, medium, high, extreme), a set of labels represented by the trapezoidal fuzzy sets shown in Figure 7.3. In the interests of simplifying notations, we represent these terms by α_i, respectively:

$$T_{DoA} = \{\alpha_1, \alpha_2, \alpha_3, \alpha_4, \alpha_5\}$$

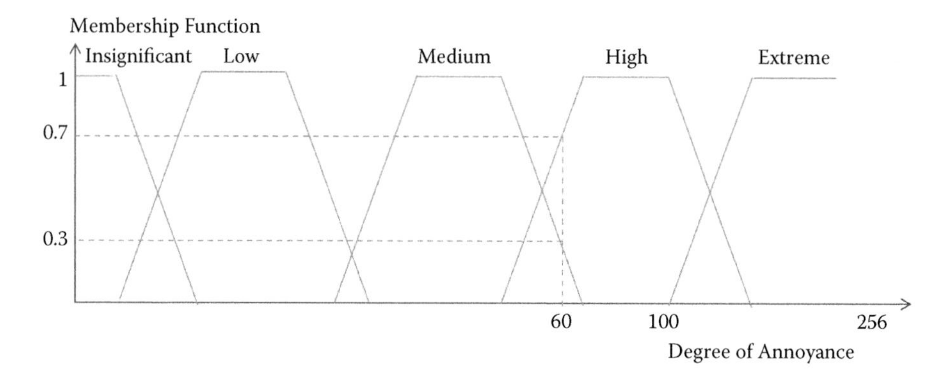

FIGURE 7.3
Fuzzy representation of degree of annoyance measure.

Let x be an imprecise value of type *DoA*. We use the following form to represent it:

$$DoA(x) = (\mu_{A_1}(x)/\alpha_1, \mu_{A_2}(x)/\alpha_2, \mu_{A_3}(x)/\alpha_3, \mu_{A_1}(x)/\alpha_4, \mu_{A_5}(x)/\alpha_5)$$

Representation of population density: To represent population density measure, we use the following linguistic variable:

$$(DoP, R^+, T_{DoP})$$

where *DoP* is the name of the variable, R^+ is the universe of discourse, and T_{DoP} = (small, medium, high), a set of labels represented by the trapezoidal fuzzy sets, as shown in Figure 7.4. In the interests of simplifying notations, we represent these terms by β_j, respectively:

$$T_{DoP} = \{\beta_1, \beta_2, \beta_3\}$$

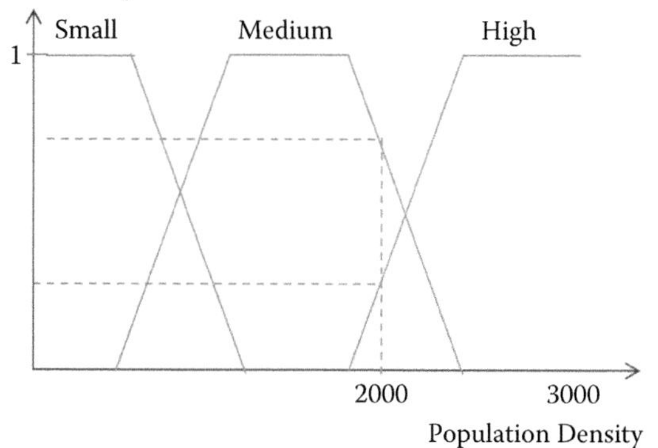

FIGURE 7.4
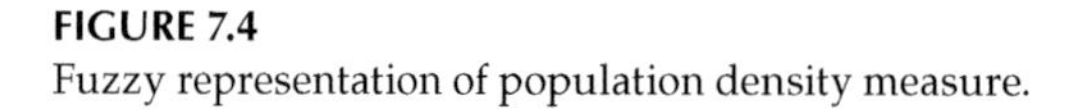
Fuzzy representation of population density measure.

Let y be an imprecise value of type *DoP*. We use the following general form to represent it:

$$DoP(t) = (\mu_{\beta_1}(y)/\beta_1, \mu_{\beta_2}(y)/\beta_2, \mu_{\beta_3}(y)/\beta_3)$$

Representation of impact of annoyance: To represent the impact of annoyance, which is the output of the combination process of impact of annoyance and population density, we use the following linguistic variable:

$$(IoA, R^{+}, T_{IoA})$$

where *DoI* is the name of the variable, R^{+} is the universe of discourse, and T_{IoA} = {very low, low, high, very high}. For the sake of simplify, we will represent these terms by γ_i, respectively:

$$IoA = (\gamma_1, \gamma_2, \gamma_3, \gamma_4\}$$

Let z be an imprecise value of type *IoA*. We use the following form to represent it:

$$IoA(z) = (\mu_{\gamma_1}(x)/\gamma_1, \mu_{\gamma_2}(x)/\gamma_2, \mu_{\gamma_3}(x)/\gamma_3, \mu_{\gamma_1}(x)/\gamma_4$$

7.3.2.2 Fuzzy Fusion

The model of fuzzy fusion that we propose allows the combination of qualitative and quantitative measures. This model is based on fuzzy inference. Based on the works of Siler and Buckeley (2005), Zadeh (1975), and Feng et al. (2010), we examine its different components.

7.3.2.2.1 Fuzzy Rule of Inference

Fuzzy inference can describe the relationship between the causal and resultant variables in terms of a collection of if-then fuzzy rules. Each fuzzy if-then rule can take the following general form:

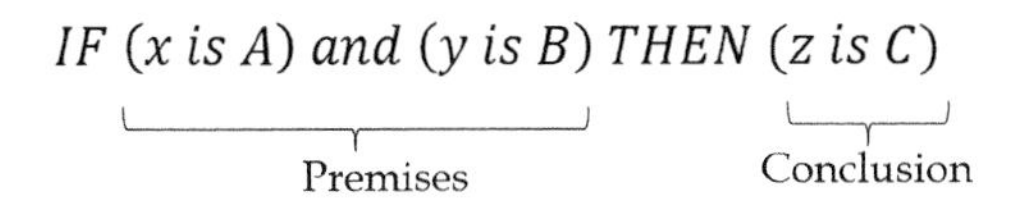

This rule corresponds to the general statement and can be presented in the form of predicates (using the notation presented earlier in Section 7.3.2) as follows:

$$A(x) \wedge B(y) \rightarrow C(z)$$

where *A*, *B*, and *C* are three linguistic variables to which the fuzzy sets α_i, β_j, and γ_k are associated, respectively.

Now, an instance of the previous rule is given by

$$A(x') \wedge B(y') \rightarrow C(z')$$

The objective is to compute the conclusion $C(z')$ given the general statement $A(x) \wedge B(y) \rightarrow C(z)$ and $A(x') \wedge B(y')$.

7.3.2.2.2 Computing Fuzzy Conclusion

To find the conclusion $C(z')$, we use the extended modus ponens rule, as defined by Zadeh (1975).

If $A(x') \wedge B(y')$
and $A(x) \wedge B(y) \rightarrow C(z) \quad (R)$
Then $C(z') = (A(x') \wedge B(y')) \circ (A(x) \wedge B(y) \rightarrow C(z'))$

where R is a fuzzy relation. Note that there are different approaches to calculate the relation (Siler and Buckeley, 2005). In our case, this relation has been established by the experts in the form of a table (see Table 7.3 for an example).

$\circ$ is a max–min or product-sum composition.
$\wedge$ is a min or product operator.

Now the membership degree function of $C(z')$ is given by

$$\mu_{\gamma_k}(z') = (\mu_{\alpha_i}(x') \wedge \mu_{\beta_i}(y')) \circ \mu_R((x,y),z))$$

$$\mu_{\gamma_k}(z') = \mu_{\alpha_i \wedge \beta_j}(x',y') \circ \mu_R((x,y),z))$$

where $\mu_{\alpha_i \wedge \beta_j}(x',y')$ and $\mu_R((x,y),z))$ are raw vectors and matrices of compatible dimensions.

TABLE 7.3

Table Used to Represent the Fuzzy Relation R

	Degree of Annoyance (DoA)				
Population Density (DoP)	α_1	α_2	α_3	α_4	α_5
β_1	γ_1	γ_1	γ_1	γ_1	γ_2
β_2	γ_1	γ_1	γ_3	γ_3	γ_3
β_3	γ_1	γ_2	γ_3	γ_4	γ_4

7.3.2.2.3 Application

The derived measure impact of annoyance is evaluated by combining population density and degree of annoyance. We can write

$$DoA(x) \wedge DoP(y) \rightarrow IoA(z)$$

The relation R corresponding to this rule is given by Table 7.3.

A simple representation of the relation R can be a binary relation given by the matrix below:

$$\mu_{R_{\alpha_i \beta_j, i=1..5, j=1...3}} = \begin{bmatrix} 1 & 0 & 0 & 0 \\ 1 & 0 & 0 & 0 \\ 1 & 0 & 0 & 0 \\ 1 & 0 & 0 & 0 \\ 1 & 0 & 0 & 0 \\ 0 & 1 & 0 & 0 \\ 1 & 0 & 0 & 0 \\ 0 & 0 & 1 & 0 \\ 0 & 0 & 1 & 0 \\ 0 & 0 & 0 & 0 \\ 0 & 0 & 1 & 0 \\ 0 & 0 & 0 & 1 \\ 0 & 1 & 0 & 0 \\ 0 & 0 & 1 & 0 \\ 0 & 0 & 0 & 1 \end{bmatrix}$$

Now given

$$DoA(x_1) = (0/\alpha_1, 0/\alpha_2, 0.3/\alpha_3, 0.7/\alpha_4, 0/\alpha_5)$$

$$DoP(y_1) = (0/\beta_1, 0.2/\beta_2, 0.8/\beta_3)$$

how should we calculate the combination of these two variables? Let be $IoA(z_1)$.

Solution:

$$\mu_{\gamma_k}(z_1) = \mu_{\alpha_i \wedge \beta_j}(x', y') \circ \mu_R((x,y),z))$$

$\mu_{\alpha_i \wedge \beta_j}(x_1, y_1)$ is given by Table 7.4.

TABLE 7.4

Table Used to Calculate $\mu_{\alpha_i \wedge \beta_j}(x_1, y_1)$

DoA	DOP	$\mu_{\alpha_i}(x_1)$	$\mu_{\beta_j}(y_1)$	$\mu_{\alpha_i \min \beta_j}(x_1, y_1)$
α_1	β_1	0	0	0
α_1	β_2	0	0.2	0
α_1	β_3	0	0.8	0
α_2	β_1	0	0	0
α_2	β_2	0	0.2	0
α_2	β_3	0	0.8	0
α_3	β_1	0.3	0	0
α_3	β_2	0.3	0.2	0.2
α_3	β_3	0.3	0.8	0.3
α_4	β_1	0.7	0	0
α_4	β_2	0.7	0.2	0.2
α_4	β_3	0.7	0.8	0.7
α_5	β_1	0	0	0
α_5	β_2	0	0.2	0
α_5	β_3	0	0.8	0

$$\mu_{\gamma_k}(z_1) = \begin{bmatrix} 1 & 0 & 0 & 0 \\ 1 & 0 & 0 & 0 \\ 1 & 0 & 0 & 0 \\ 1 & 0 & 0 & 0 \\ 1 & 0 & 0 & 0 \\ 0 & 1 & 0 & 0 \\ 1 & 0 & 0 & 0 \\ 0 & 0 & 1 & 0 \\ 0 & 0 & 1 & 0 \\ 0 & 0 & 0 & 0 \\ 0 & 0 & 1 & 0 \\ 0 & 0 & 0 & 1 \\ 0 & 1 & 0 & 0 \\ 0 & 0 & 1 & 0 \\ 0 & 0 & 0 & 1 \end{bmatrix}$$

Thus, using max-min composition, we obtain

$$\mu_{\gamma_k}(z_1) = [0, 0, 0.3, 0.7]$$

So in fuzzy set form, $IoA(z_1) = (0/\gamma_1, 0/\gamma_2, 0.3/\gamma_3, 0.7/\gamma_4)$.

7.4 Prototype and Experimentation

Our objective is to assess the feasibility, effectiveness, and expressivity of our model within an off-the-shelf database management system (DBMS). The global architecture of our prototype consists of three main components (Figure 7.5):

1. The DBMS for managing the fact and dimension tables of annoyance
2. The OLAP server for the multidimensional analysis of annoyance, which consists of the multidimensional data cubes constructing and exploitation
3. The user interface, which allows visualizing the annoyance aggregation data using tabular or cartographic representation

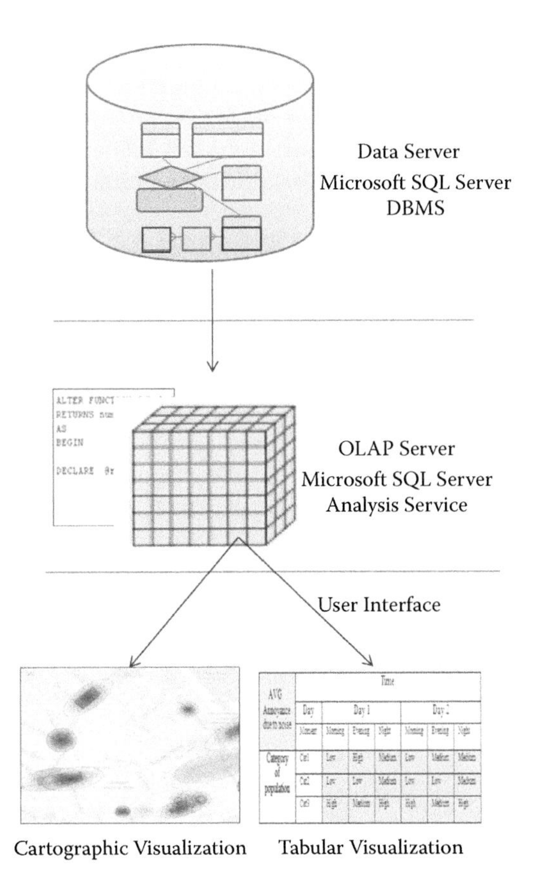

FIGURE 7.5
System architecture.

7.4.1 Implementation of Annoyance Data Warehouse

To implement the annoyance data warehouse, we have opted for a relational on line analytical processing (ROLAP) solution (Malinowski and Zimányi, 2008) using Microsoft SQL Server. Note that the principle of this implementation is not DBMS dependent. After establishing the conceptual schema and creating the database, the next step is the data integration and loading. This step is usually the most complex and critical in the process of data warehousing. To feed the annoyance data warehouse, we have used the SQL Server Integration Services (SSIS) tool. This is a service of the Business Intelligence Development Studio solution that is responsible for the extraction, transformation, and loading of data, commonly called ETL (Kimball and Ross, 2002).

7.4.2 Implementation of Fuzzy Fusion in Data Warehouse

We have implemented fuzzy fusion using user-defined functions. This function is used to calculate the derived measure impact of annoyance. Most modern DBMSs allow the extensibility of their typing system and functions. Our choice was Microsoft SQL Server. We show that using user-defined functions allows us to develop our model. It can be used in any SQL query. The only limitation is that those functions cannot be used in the SSAS OLAP tool for interactive analysis of the multidimensional model.

```
Function Fuzzy-Fusion (x, y, Op1, Op2)
Input:
x, y are two values of imprecise nature // for instance:
population density and typical individual annoyance.
Op1: the premises combination.
Op2: the composition operator.
Begin
Construct the implication vector
Read the relation matrix
Do the composition
  Return the result            // annoyance impact
End
```

This function has been integrated within the data server (Figure 7.5).

7.5 Conclusion and Future Work

The main objective of this work is the extension of conventional data warehouses to allow the integration and processing of both qualitative and quantitative measures. Based on computing with words methodology, we had, in a previous work (Amanzougarene et al., 2012), introduced qualitative measures and aggregates as an extension of the multidimensional data model of a data warehouse. Using these measures and aggregates, OLAP queries allow the decision maker to manipulate data in a qualitative fashion using linguistic terms. This chapter has proposed an extension to deal with fuzzy data model that is in intermediary solution between conventional numeric measures and crisp qualitative values. To illustrate our proposal, we have considered the case of urban building site annoyance. We have proposed an original approach that allows managing the annoyance and its inherent imprecision, as in commonsense reasoning, by using linguistic expressions.

In our future work, we will extend our approach of evaluation in order to include spatial and temporal extent of annoyance as measures in the multidimensional data model of the data warehouse. We will also define aggregation operations allowing data processing of these extents (e.g., fusion of annoyance influence areas and concatenation of exposure time interval). Indeed, that will improve the decisions of managers of public spaces concerning urban building site planning. So far, we have focused on the measurement and aggregation in a qualitative fashion for the annoyance from previous inputs. We also intend to extend our approach to the prediction of annoyance, so that it helps predict the best place and time for a new building site.

References

Amanzougarene, F., Chachoua, M., Zeitouni, K. 2012. Qualitative representation of building sites annoyance. In *Proceedings of the 2012 ACM Workshop on City Data Management*, pp. 13–20, Maui, Hawaii.

Burdick, D., Deshpande, P. M., Jayram, T. S., Ramakrishnan, R., Vaithyanathan, S. 2007. OLAP over uncertain and imprecise data. *VLDB Journal*, 16(1), 123–144.

Delgado, M., Molina, C., Ariza, L. R., Sánchez, D., Miranda M. A. V. 2007. F-cube factory: A fuzzy OLAP system for supporting imprecision. *International Journal of Uncertainty, Fuzziness and Knowledge-Based Systems (IJUFKS)*, 15, 59–81.

Feng, F., Li, Y., Li, C., Han, B. 2010. *Soft Set Based Approximate Reasoning: A Quantitative Logic Approach*. Springer, Berlin, pp. 245–255.

Guski, R., Felscher-suhr U. 1999. The concept of noise annoyance: How international experts see it. *Journal of Sound and Vibration*, 223(4): 513–527.

Inmon, W. H. 2005. *Building the Data Warehouse*. 4th ed. John Wiley & Sons, New York.

Kimball, R., Ross, M. 2002. *The Data Warehouse Toolkit: The Complete Guide to Dimensional Modeling*. 2nd ed. John Wiley & Sons, New York.

Laurent, A. 2001. Generating fuzzy summaries from fuzzy multidimensional databases. In *Proceedings of the 4th International Conference on Advances in Intelligent Data Analysis*, pp. 24–33, Cascais, Portugal.

Malinowski, E., Zimányi, E. 2008. *Advanced Data Warehouse Design: From Conventional to Spatial and Temporal Applications*. Springer-Verlag, Berlin.

Molina, C., Rodríguez-Ariza, L., Sánchez, D., Vila, M. A. 2006. A new fuzzy multidimensional model. *IEEE Transactions on Fuzzy Systems*, 14(6), 987–912.

Moser, G., Robin, M. 2006. Environmental annoyances: An urban-specific threat to quality of life? *European Review of Applied Psychology*, 56(1), 35–41.

Nordin, S., Lidén, E. 2006. Environmental odor annoyance from air pollution from steel industry and bio-fuel processing. *Journal of Environmental Psychology*, 26(2), 141–145.

Robin, M., Matheau-Police, A., Couty, C. 2007. Development of a scale of perceived environmental annoyances in urban settings. *Journal of Environmental Psychology*, 27(1), 55–68.

Siler, W., Buckeley, J. 2005. *Fuzzy Expert Systems and Fuzzy Reasoning*. John Wiley & Sons, Hoboken, NJ.

Yager, R. R. 2007. Aggregation of ordinal information. *Fuzzy Optimization and Decision Making*, 6(3), 199–219.

Zadeh, L. A. 1975. The concept of a linguistic variable and its application to approximate reasoning. *Information Sciences*, 8(3), 199–249.

8

Quality Assessment in River Network Generalization by Preserving the Drainage Pattern

L. Zhang[1,2] **and E. Guilbert**[3]

[1]*Key Laboratory of Virtual Geographic Environment, Ministry of Education, Nanjing Normal University, Nanjing, People's Republic of China*

[2]*Jiangsu Center for Collaborative Innovation in Geographical Information Resource Development and Application, Nanjing, People's Republic of China*

[3]*Department of Geomatic Sciences, Laval University, Quebec, Canada*

CONTENTS

ABSTRACT The drainage pattern of a river network is the arrangement in which a stream erodes the channels of its network of tributaries. It can reflect the geographical characteristics of a river network to a certain extent because it depends on the topography and geology of the land. Whether in cartography or geographic information system (GIS), hydrography is one of the most important feature classes to generalize in order to produce representations at various levels of detail. Cartographic generalization is an intricate process whereby information is selected and represented on a map at a certain scale not necessarily preserving all geographical or other cartographic details. There are many methods for river network generalization, but the generalized results are always inspected by expert cartographers visually. This chapter proposes a method that evaluates the quality of a river network generalization by assessing if drainage patterns are preserved. This method provides a quantitative value that estimates the membership of a river network in different drainage patterns. A set of geometric indicators describing each pattern are presented, and the membership of a network is defined based on fuzzy logic. For each pattern, the fuzzy set membership is given by a defined IF-THEN rule composed of several indicators and logical operators. Assessing the quality of a generalization is done by comparing and analyzing the values before and after the network generalization. This assessment method is tested with several river network generalization methods on different sets of networks, and results are analyzed and discussed.

8.1 Introduction

Automated map generalization is an important issue and a major challenge in cartography and GIS research. Regarded as the skeleton of the terrain, the drainage system should be considered first in automated map generalization. In a drainage system, a stream or a river is a natural watercourse, usually freshwater, flowing toward an ocean, a lake, or another river. Apart from a few cases where a river simply flows into the ground or dries up completely before reaching another body of water, rivers always connect together to form networks, achieving a particular drainage pattern. Further, as the most important component of the drainage system, generalization of rivers properly becomes a focal point because (1) rivers are an important part of the land and need to be represented in maps of any kind, and (2) rivers are fundamental concepts used for various analyses in geoscience. For instance, geologists can get original slope and original structure from drainage patterns. As a set of line features, river networks are generalized from a large scale to a small scale in two main steps: selective omission and selected tributary simplification.[1] There are many methods for selective omission and simplification of tributaries for rivers selected as individual line features, but while most research focuses on river networks during the generalization, generalized results are always inspected by expert cartographers visually. Considering map generalization from the geographic level first,[2,3] drainage patterns can be applied in river network generalization, as patterns are important in generalization and should be explicitly measured and evaluated.[4]

At present, many researchers have started to pay attention to geographical features of river networks during the process of generalization,[2,5,6] which follows the idea that "generalization is not a mere reduction of information—the challenge is one of preserving the geographic meaning."[7] Considering drainage pattern as a geographical factor in river network generalization helps us to retain geographical features of the networks. There are several types of drainage patterns. They are commonly classified as dendritic, parallel, trellis, rectangular, radial, centripetal, and reticulate patterns.[8] Dendritic patterns, also named tree-like patterns, can usually be found where there is no strong geological control.[9] Parallel, trellis, and rectangular drainage patterns develop in areas with strong regional slopes but have their own specific characteristics. Streams radiating from a high central area form a pattern of radial drainage, while streams forming a centripetal one gather in low-lying land. Reticulate drainage patterns are usually found on floodplains and deltas where rivers often interlace with one another.[10] The first three drainage patterns are illustrated in Figure 8.1.

From the literature, little research drew attention to the aspect of the assessment of generalization.[11–14] Traditionally, generalization is evaluated by visual assessment by the cartographic experts to grade the quality by questionnaires.[15] This method is based on knowledge and experience, and it is rather subjective.[12,16] For the quality assessment of river network generalization, especially in selective omission, little work is available. The most relevant one is the coefficient of line correspondence (CLC) calculated to compare the generalized data with existing data.[5,6] CLC is given based on length only, which cannot assess the generalized river network comprehensively. Overall, the methods evaluating river network generalization quality are not well developed, and visual assessment is still often used.

In this chapter, drainage patterns are applied to evaluate a generalized river network. Fuzzy logic is adopted in the method, where several indicators are used to quantify the features of drainage patterns. In this work,

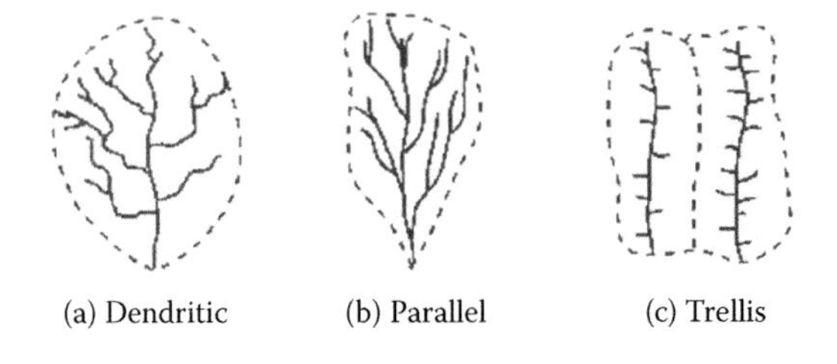

FIGURE 8.1
Some examples of drainage network pattern. The dashed line is the catchment of each river network. (Diagrams are modified from Ritter, M.E., The physical environment: An introduction to physical geography, 2006, http://www.earthonlinemedia.com/ebooks/tpe_3e/title_page.html.)

the dendritic, parallel, and trellis patterns are addressed. In addition, this research focuses on the selective omission in river network generalization, as feature selection is the first step in any generalization. In Section 8.2, the fundament of fuzzy logic is introduced. Section 8.3 provides the details of the evaluation method. Finally, the method is implemented and tested.

8.2 Fundamentals of Fuzzy Logic

Zadeh (1965) introduced the fuzzy set theory and fuzzy logic in 1965.[17] Reasoning on fuzzy sets is based on fuzzy operators and fuzzy rules.[18]

8.2.1 Fuzzy Set

A fuzzy set is a set whose membership is not defined by a binary value (an element belongs or not to a set), but by a value between 0 and 1 corresponding to different grades of membership. A membership function (MF) associated with a given fuzzy set maps an input value to its appropriate membership value. There are five common membership functions in use, as shown in Figure 8.2. Fuzzy set theory allows approximated reasoning on values that are imprecise or incomplete.

In this work, three of the MFs shown in Figure 8.2 are used: Gaussian, Z curve, and S curve MFs. They can be described mathematically as follows.

The Gaussian function is given as

$$g(x;\sigma,\mu) = e^{\frac{-(x-\mu)^2}{2\sigma^2}} \tag{8.1}$$

where x is input, μ is the mean, and σ is the standard deviation controlling the width of the curve.

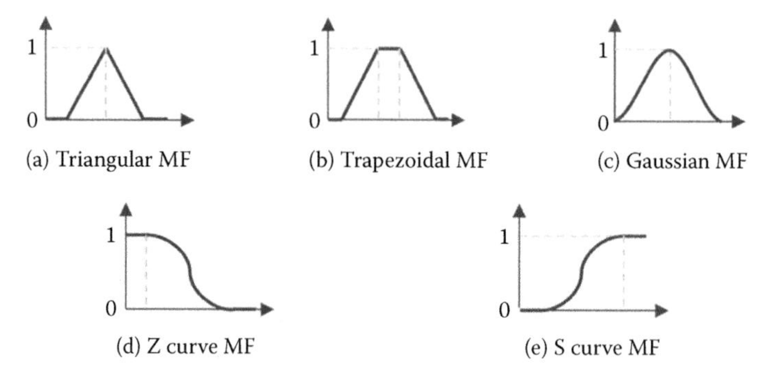

FIGURE 8.2
Five common MFs: (a) triangular MF, (b) trapezoidal MF, (c) Gaussian MF, (d) Z curve MF, and (e) S curve MF. MFs (c) through (e) are used in this study.

The Z curve MF is a spline-based function of x:

$$z(x;a,b)=\begin{cases} 1, & x \le a \\ 1-2\left(\dfrac{x-a}{b-a}\right)^2, & a < x \le \dfrac{a+b}{2} \\ 2\left(\dfrac{x-b}{b-a}\right)^2, & \dfrac{a+b}{2} < x < b \\ 0, & x \ge b \end{cases} \tag{8.2}$$

where a and b are the extremes of the sloped portion of the curve, and $a < b$.

The S curve MF is a mirror-image function of the Z curve:

$$s(x;a,b)=\begin{cases} 0, & x \le a \\ 2\left(\dfrac{x-a}{b-a}\right)^2, & a < x \le \dfrac{a+b}{2} \\ 1-2\left(\dfrac{x-b}{b-a}\right)^2, & \dfrac{a+b}{2} < x < b \\ 1, & x \ge b \end{cases} \tag{8.3}$$

where a and b are the extremes of the sloped portion of the curve, and $a < b$.

8.2.2 Fuzzy Operator

In fuzzy logic, the truth of any statement is a matter of degree between 0 and 1. Zadeh (1965) suggested the minimum, maximum, and complement methods for AND, OR, and NOT operators, respectively. For two fuzzy set values A and B within the range (0, 1), fuzzy logic operations are (Figure 8.3)

$$\begin{gathered} \text{A AND B} = \min(\text{A, B}) \\ \text{A OR B} = \max(\text{A, B}) \\ \text{NOT A} = 1 - \text{A} \end{gathered} \tag{8.4}$$

8.2.3 Fuzzy Rule

Fuzzy rules, also called IF-THEN rules, are used to represent the conditional statements with fuzzy sets and fuzzy operators. A single fuzzy IF-THEN rule is like

IF (a is X) THEN (b is Z)

where a is the input variable, b is the output, and X and Z are defined by fuzzy sets. Here, it should be noted that there is no ELSE part in a fuzzy rule.

Usually, there are several fuzzy rules for a fuzzy logic application. In the process, all rules should be evaluated, and the outputs must be combined in

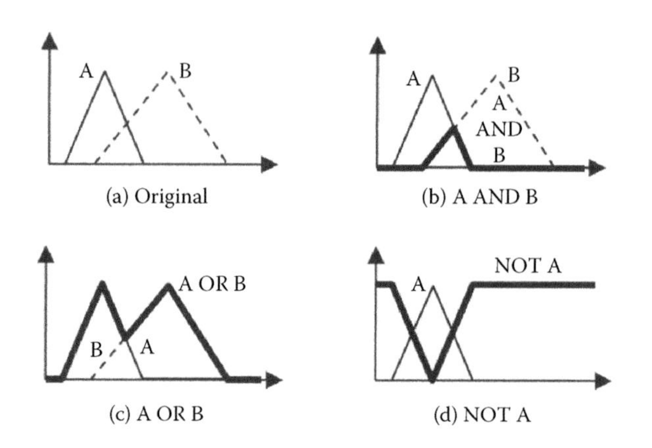

FIGURE 8.3
AND, OR, and NOT operations. AND is the intersection operation, OR is the union operation, and NOT is the complement operation.

some way in order to get a final result, which is called aggregation. Especially for a fuzzy control system, defuzzification is typically needed, which is a process of producing a quantifiable result.[19]

8.3 Evaluation Methods

8.3.1 Predicates and Fuzzy Rules

In this research, several indicators listed in Table 8.1 are used to classify networks into drainage patterns and evaluate them. These indicators describe the characteristics of each pattern and have been identified by Zhang and Guilbert:[20] average junction angle (α), bended tributary percentage (β), average length ratio (γ), and catchment elongation (δ). All these values are calculated from a river network.

Here, for each drainage pattern, a rule is defined to predict the pattern of a river network:[20]

1. IF (α IS acute) AND (δ IS broad), THEN pattern IS dendritic.
2. IF (α IS very acute) AND NOT (β IS bended) AND (γ IS long) AND (δ IS elongated), THEN pattern IS parallel.
3. IF (α IS right) AND NOT (β IS bended) AND (γ IS short) AND (δ IS elongated), THEN pattern IS trellis.

Therefore, eight predicates are formulated into membership functions. They are list Table 8.2.

TABLE 8.1

List of Geometric Indicators

Indicator	Description	Illustration
Junction angle	The angle is composed by upper tributaries	
Sinuosity	A ratio of the channel length to the valley length[21]	
Length ratio	A ratio of the tributary length to the main stream length	
Catchment elongation	A ratio of the depth to the breadth of a catchment	

TABLE 8.2

Predicates in Fuzzy Rules

Predicates	MFs	Examples
α IS acute	$z(x;a,b)$	$z(\alpha;45°,90°)$
α IS very acute		$z(\alpha;30°,60°)$
γ IS short		$z(\gamma;0,1)$
δ IS broad		$z(\delta;1,3)$
α IS right	$g(x;a,b)$	$g(\alpha;10°,90°)$
β IS bended	$s(x;a,b)$	$s(\beta;0,1)$
γ IS long		$s(\gamma;0,1)$
δ IS elongated		$s(\delta;1,3)$

8.3.2 Evaluation Method for Each Pattern

From the fuzzy rule and MFs introduced in Section 8.3.1, the evaluation method for each pattern can be given by integrating fuzzy operators. The detailed formulas for evaluating each pattern can be described as follows.

8.3.2.1 Dendritic Pattern

The degree of a dendritic network can be calculated as "(α IS acute) AND (δ IS broad)," which can be represented in the following formula:

$$f(\alpha,\delta) = \min(z(\alpha;a,b), z(\delta;a',b')) \tag{8.5}$$

where α and δ are inputs, and $z(\alpha;a,b)$ and $z(\delta;a',b')$ are defined MFs for an acute angle and a broad catchment, respectively.

8.3.2.2 Parallel Pattern

The degree of a parallel network is given as

$$\begin{aligned} f(\alpha,\beta,\gamma,\delta) = \min\big(&z(\alpha;a,b),\\ &1-s(\beta;a',b'),\\ &s(\gamma;a'',b''),\\ &s(\delta;a''',b''')\big) \end{aligned} \tag{8.6}$$

where α, β, γ, and δ are inputs, and $z(\alpha;a,b)$, $s(\beta;a',b')$, $s(\gamma;a'',b'')$, and $s(\delta;a''',b''')$ are MFs for a very acute angle, a bended tributary, a long tributary, and an elongated catchment, respectively.

8.3.2.3 Trellis Pattern

For the trellis pattern, the degree can be calculated as

$$\begin{aligned} f(\alpha,\beta,\gamma,\delta) = \min\big(&g(\alpha;a,b),\\ &1-s(\beta;a',b'),\\ &z(\gamma;a'',b''),\\ &s(\delta;a''',b''')\big) \end{aligned} \tag{8.7}$$

where α, β, γ, and δ are inputs, and $g(\alpha;a,b)$, $s(\beta;a',b')$, $z(\gamma;a'',b'')$, and $s(\delta;a''',b''')$ are MFs for a right angle, a bended tributary, a short tributary, and an elongated catchment, respectively.

8.4 Experiments and Results

8.4.1 Experiment Design

8.4.1.1 Strategies to Be Tested

In the tributary selection problem, there are two key issues involved: (1) How many river tributaries are selected? (2) Which ones are selected?

For the first issue, a classical principle of selection,[22] which is the so-called radical law, was enunciated by F. Topfer in 1961. The method is described as follows:

$$n_f = n_a \sqrt{\frac{M_a}{M_f}} \tag{8.8}$$

where n_f is the number of objects shown at the smaller scale M_f, and n_a is the number of objects shown at the larger scale M_a.

For the second issue, in the experiment, in order to assess whether a river network preserved its drainage pattern after generalization to a smaller scale, three generalization strategies are tested (Table 8.3). The first two are automated methods based on the hierarchical structure of the network, and the last one is the manual generalization.

In the first strategy, the Horton–Strahler scheme assigns order 1 to all branchless tributaries and higher order to those receiving tributaries following the river flow direction (Figure 8.4a). In this strategy, river tributaries are selected first based on order, and then shorter tributaries at lowest order are eliminated first (Figure 8.4b).

The second strategy is proposed by Ai et al.[2] Instead of using geometric parameters, the importance of a channel is defined by the area of its catchment area. An example of hierarchical watershed partitioning is shown in Figure 8.5. In this strategy, tributaries with small catchment areas are removed first.

TABLE 8.3

Testing on Three Strategies

No.	Strategies
1	Horton–Strahler order + length[23]
2	Watershed partitioning[2]
3	Generalized manually

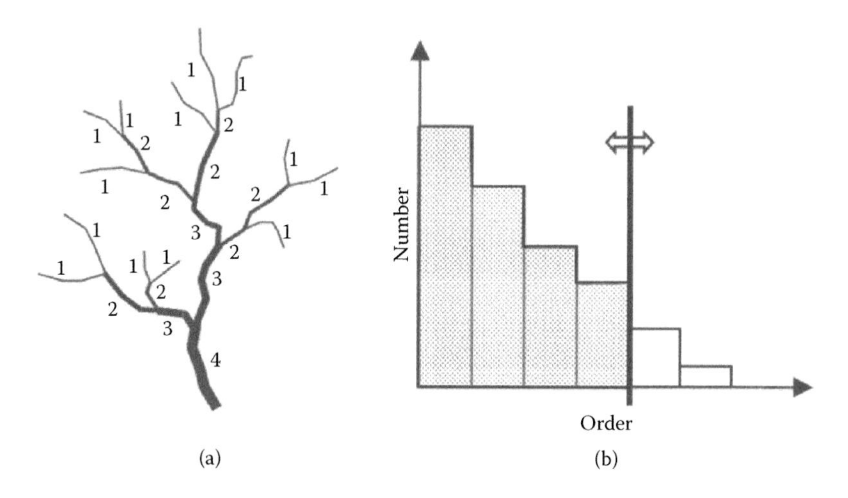

FIGURE 8.4
(a) Horton–Strahler ordering. (Modified from Li, Z., *Algorithmic Foundation of Multi-Scale Spatial Representation*, CRC Press, Boca Raton, FL, 2007.) (b) Strategy where lower-order streams are removed.

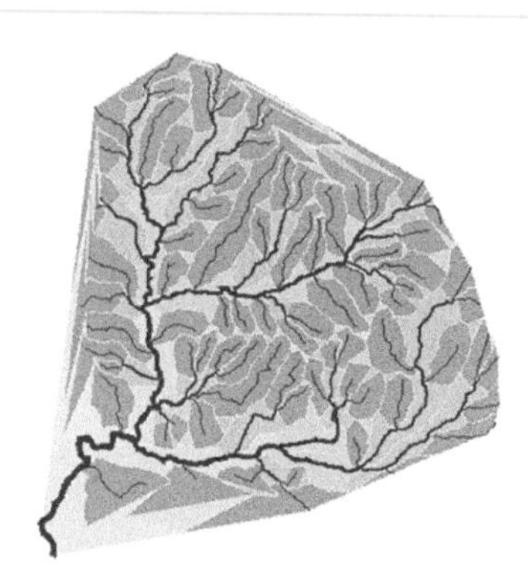

FIGURE 8.5
Hierarchical partitioning of river catchments. The darker the catchment, the lower the order.

8.4.1.2 Testing Data

Russian River data sets are tested in the experiment. Two different scales are used in the experiments: 1:24,000 scale (24K) and 1:100,000 scale (100K). The small-scale data set is provided by the U.S. National Hydrography Dataset (NHD). From the history of the establishment of the NHD, the medium-resolution data set is built first, and then a conflation tool is used to generate the 24 K hydrological data set from the medium-resolution data. The medium resolution in the NHD data is at 100 K. Therefore, the 100 K data are not generalized from 24 K, but built manually. The Horton–Strahler order scheme was then computed.

8.4.2 Case Studies in Russian River

1. Case 1: A first tested river network is illustrated in Figure 8.6a. It is a dendritic network with a membership value of 0.933. Figure 8.6b–d shows the result of generalization strategies 1–3, respectively.

   In Figure 8.6, all generalized results are good, but (d) is better than (b) and (c) by visual checking because (d) looks more balanced. River network (b) has more short tributaries due to the sequencing of the method. In strategy 1, lower-order tributaries have been removed first, although they are long; some higher-order but short tributaries remain.

   Table 8.4 provides the assessment result of the first case study, and it shows the same findings with the visual assessment. All generalized river networks preserve the dendritic pattern, and the

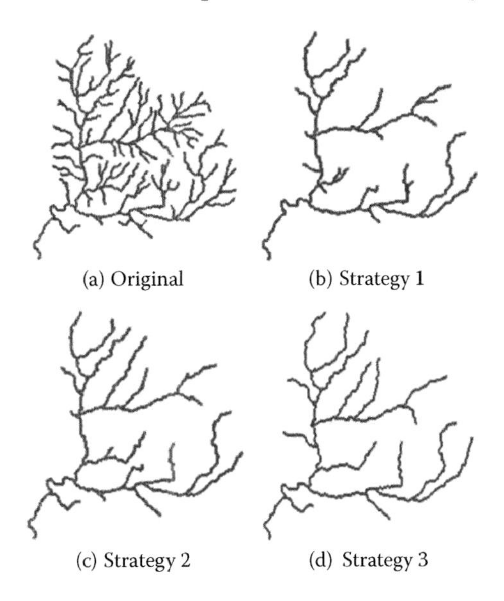

FIGURE 8.6
Generalized river network for dendritic case. (a) Original dentritic network. (b–d) Generalized networks.

TABLE 8.4
Assessment Result of Dendritic Case

	Indicator				Membership Value		
No.	α	β	γ	δ	**D**	**P**	**T**
	53.24°	3.68%	0.69	1.14	**0.933**	0.010	0.001
1	61.56°	2.70%	0.71	1.16	**0.729**	0	0.013
2	61.52°	8.57%	0.59	1.16	**0.730**	0	0.013
3	59.47°	8.11%	0.64	0.99	**0.793**	0	0

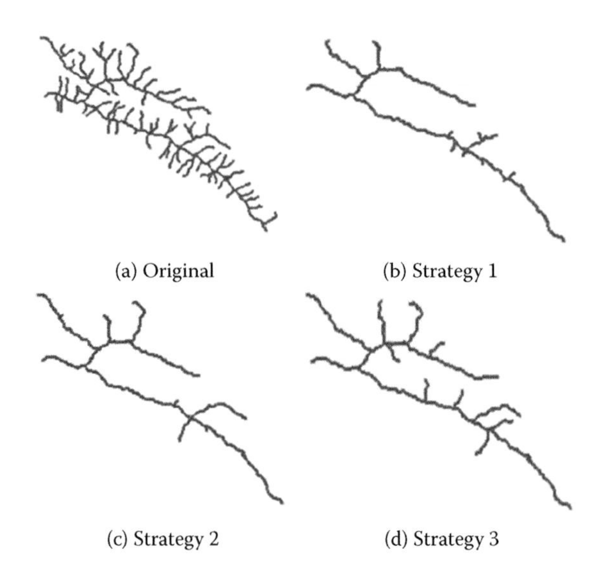

FIGURE 8.7
Generalized river network for trellis case. (a) Original trellis network. (b–d) Generalized networks.

membership values are 0.729, 0.730, and 0.793 for strategies 1, 2, and 3, respectively. The membership value (0.793) of the river network generalized manually is greater than other results. Strategy 3 brings a better result in this case study.

2. Case 2: The second case study is in Figure 8.7a. This river network is also a portion of the Russian River, which is arranged in a trellis.

   Although it is not a satisfying result, the network in Figure 8.7b has more features of the trellis pattern due to the short preserved tributaries. The generalized network in Figure 8.7c has more long tributaries because they have larger catchments. The manually generalized network (d) is good by visual assessment.

   In Table 8.5, from the assessment result, all results preserved the trellis pattern. Strategy 2 gets a greater membership (0.677) to the

TABLE 8.5
Assessment Result of Trellis Case

	Indicator				Membership Value		
No.	α	β	γ	δ	D	P	T
	81.14°	1.49%	0.20	3.17	0	0	**0.675**
1	100.68°	3.33%	0.18	3.29	0	0	**0.565**
2	98.83°	3.85%	0.25	3.09	0	0	**0.677**
3	86.38°	4.35%	0.28	3.65	0	0	**0.842**

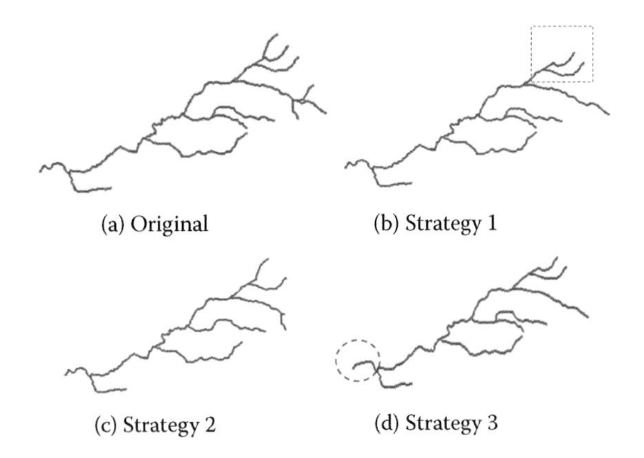

FIGURE 8.8
Generalized river network for parallel case. (a) Original parallel network. (b–d) Generalized networks.

trellis pattern than strategy 1 (0.565). The manually generalized river network is better than the others because its membership value is 0.842, which is the greatest of all generalized networks.

3. Case 3: The river network tested in this experiment is a parallel river network. It is shown in Figure 8.8a.

   In Figure 8.8, visually, the shape of network (b) is more similar to a parallel network because the tributaries in the dashed box of network (b) provide a better general representation than network (c). Networks (b) and (d) are almost the same; the only difference is that some tributaries in network (d) have been smoothed.

   From Table 8.6, memberships of each strategy are 0.929, 0.632, and 0.926, respectively. Strategy 1 provides a better result because the membership value (0.929) is greater than those of the others, but it is nearly the same as 0.926 from strategy 3. The possible reason is the data from the NHD are smoother, such as the tributary in the dashed circle in Figure 8.8d. This result is confirmed by visual assessment.

TABLE 8.6
Assessment Result of Parallel Case

	Indicator				Membership Value		
No.	α	β	γ	δ	D	P	T
	39.67°	10.53%	0.69	3.14	0	**0.792**	0
1	34.02°	15.38%	0.81	3.86	0	**0.929**	0
2	41.15°	15.38%	0.57	3.43	0	**0.632**	0
3	27.63°	0%	0.81	4.84	0	**0.926**	0

8.5 Conclusions

From the experiment results, several conclusions can be given as follows:

1. In general, the evaluation method based on the membership degree of a fuzzy rule for a drainage pattern is useful. From a large scale to a small scale, to a generalized river network, the drainage pattern is better preserved if the membership value is high.
2. By evaluating generalized river networks from the point of drainage patterns, networks generalized manually are always with high membership values and preserve a good drainage pattern. A good generalized result does not depend on only one or two factors; many factors, such as tributary spacing and balance, are involved in the manual generalization process.
3. One limitation of the proposed evaluation method is that it focuses on the drainage pattern only. Some other aspects simply cannot be assessed by the membership value.
4. Another limitation is that the validation is based on case studies, and a comparison shall be conducted on the whole network.

A first direction for future work is therefore to conduct larger sets of tests on whole networks in different types of terrain to perform a statistical comparison and get a more robust evaluation. Further, other drainage patterns, such as reticulate, radial, and centripetal patterns, should be characterized and considered in the quality assessment. Finally, drainages are considered a network formed by a river and its tributaries, but the pattern may evolve between higher and lower reaches and may require a segmentation of the river in different parts; hence, a specific approach to the partitioning of subnetworks shall be developed.

References

1. Li Z. *Algorithmic Foundation of Multi-Scale Spatial Representation*. Boca Raton, FL: CRC Press, 2007.
2. Ai T, Liu Y, Chen J. The hierarchical watershed partitioning and data simplification of river network. In Riedl A, Kainz W, Elmes GA, eds., *Progress in Spatial Data Handling*. Berlin: Springer, 2006, pp. 617–632.
3. Poorten PM Van Der, Jones CB. Characterisation and generalisation of cartographic lines using Delaunay triangulation. *Int J Geogr Inf Sci* 2002;16(8):773–794.

4. Mackaness W, Edwards G. The importance of modelling pattern and structure in automated map generalisation. Presented at Joint Workshop on Multi-Scale Representations of Spatial Data, Ottawa, 2002.
5. Buttenfield BP, Stanislawski LV, Brewer CA. Multiscale representations of water: Tailoring generalization sequences to specific physiographic regimes. In *Proceedings of GIScience 2010*, 2010, pp. 14–17.
6. Stanislawski LV. Feature pruning by upstream drainage area to support automated generalization of the United States National Hydrography Dataset. *Comput Environ Urban Syst*. 2009;33(5):325–333.
7. Bard S, Ruas A. Why and how evaluating generalised data? In *Developments in Spatial Data Handling*. Berlin: Springer, 2005.
8. Ritter ME. The physical environment: An introduction to physical geography. 2006. http://www.earthonlinemedia.com/ebooks/tpe_3e/title_page.html.
9. Charlton R. *Fundamentals of Fluvial Geomorphology*. Psychology Press, 2008.
10. Fagan SD, Nanson GC. The morphology and formation of floodplain-surface channels, Cooper Creek, Australia. *Geomorphology* 2004;60(1):107–126.
11. Bard S. Quality assessment of cartographic generalisation. *Trans GIS* 2004;8(1):63–81.
12. Joao E. *Causes and Consequences of Map Generalisation*. London: Taylor & Francis, 1998.
13. Muller J, Weibel R, Lagrange J, Salge F. Generalization: State of the art and issues. In *GIS and Generalization: Methodology and Practice*. London: Taylor & Francis, 1995, pp. 3–17.
14. Weibel R, Dutton G. Generalising spatial data and dealing with multiple representations. In Longley P, Goodchild M, Maguire D, Rhind D, eds., *Geographical Information Systems: Principles, Techniques, Management and Applications*. 1999, pp. 125–155.
15. Weibel R. Three essential building blocks for automated generalization. In *GIS and Generalization: Methodology and Practice*. London: Taylor & Francis, 1995, pp. 56–69.
16. Mackaness W, Ruas A. Evaluation in the map generalisation process. In *Generalisation of Geographic Information: Cartographic Modelling and Applications*. Amsterdam: Elsevier, 2007, pp. 89–111.
17. Zadeh LA. Fuzzy sets. *Inf Control* 1965;8(3):338–353.
18. Bandemer H, Gottwald S. *Fuzzy Sets, Fuzzy Logic, Fuzzy Methods: With Applications*. New York: J. Wiley, 1995.
19. Leekwijck W Van, Kerre EE. Defuzzification: Criteria and classification. *Fuzzy Sets Syst* 1999;108(2):159–178.
20. Zhang L, Guilbert E. Automatic drainage pattern recognition in river networks. *Int J Geogr Inf Sci* 2013;27(12):2319–2342.
21. Schumm SA. *The Fluvial System*. New York: Wiley, 1977.
22. Topfer F, Pillewizer W. The principles of selection. *Cartogr J* 1966;3(1):10–16.
23. Rusak Mazur E, Castner HW. Horton's ordering scheme and the generalisation of river networks. *Cartogr J* 1990;27(2):104–112.

Section III

Quality Control for Spatial Products

9

Quality Control of DLG and Map Products

Pei Wang,[1] Zhiyong Lv,[2] Libin Zhao,[1] and Xincheng Guo[3]

[1]*First Institute of Photogrammetry and Remotes Sensing, Xi'an, People's Republic of China*

[2]*School of Computer Science and Engineering, Xi'an University of Technology, Xi'an, People's Republic of China*

[3]*Chang'an University, Xi'an, People's Republic of China*

CONTENTS

ABSTRACT The product of a digital line graph and map (DLGM) is a vector data set of a map that is software independent. The advantages of this product, such as commonality and multipurpose applications, have been confirmed by the mapping industry and geographic information system (GIS) applications. First, this study outlines the technological framework of DLGM products. Second, quality control methods and mechanisms are described, which involve dynamic modelling based on a topological map model, data dictionaries, and the DLGM integration template. Finally, the role of DLGM quality control is discussed. In recent years, a number of provincial fundamental DLGM products have been produced by our DLGM integrated technology system. The results of applications have proved that the product quality control theory and technology adopted by our system are effective and feasible.

9.1 Introduction

A digital line graph and map (DLGM) is a vector data set of a topographic map that stores geometric features, the spatial relationship among geographic features, corresponding features' attributes, and symbols.[1] We have been studying the integration of geographic databases and maps for many years and have obtained abundant information on integrated manufacturing systems. The production of DLGMs is a main achievement of our research. An integrated manufacturing system, which is introduced in the following paragraphs, is a system that produces DLGMs.

9.1.1 Topological Map Model

A topological map model is a physical model of a DLGM product. In this model, a feature (entity) consists of its geometry, attributes, and symbol. The components of a feature are equal in a topographic map model. In other words, there is no subordinate relationship between each component. This characteristic comprises the main difference between a topological map model and a traditional map model, such as a geographic information system (GIS) model.

9.1.2 MapStore Platform

MapStore, which is based on a topographic map model, is the modelling software platform for DLGMs. This platform can deal with the complicated spatial topological relationships between features and also realize dynamic modelling and provide accurate map products through the integration of a

DLG and map data. The functions of this platform include defining a model, data modelling and editing, quality inspection, and data import and export.

9.1.3 Integration Process

It is difficult to model and autopreserve the topological relationships between each feature of the fundamental geographic information product. To solve this problem, the integration process includes a series of methods (e.g., the geometric construction method). Therefore, DLGM products can be produced from data and maps with higher quality and efficiency.

9.1.4 Quality Control

Quality control is conducted according to the quality standards of spatial data.[2] This process ensures that DLGM products have the ability to complete quality control methods. Each link in the production system is a quality control unit, and the links coordinate with one another.

Compared with other traditional spatial maps, the DLGM is a new digital map product, and its basic theory and technical methods have attracted the attention of the whole industry. In this chapter, the principle and methods of DLGM quality control are discussed in detail.

9.2 Consistency Quality Control Mechanisms of DLGM

9.2.1 DLGM Integration Model

The integration model of a DLGM is the most essential mechanism that guarantees the consistent quality of DLGM products. This model is a topological map model, as mentioned in Section 9.1.1. Figure 9.1 shows its structure.

A completed entity is made up of three subordinate elements: a feature's geometry, attributes, and symbol. The entity and its subordinate elements form a ternary tree relationship. Just as a father node manages its child nodes, the entity element manages a feature's geometry, attributes, and map

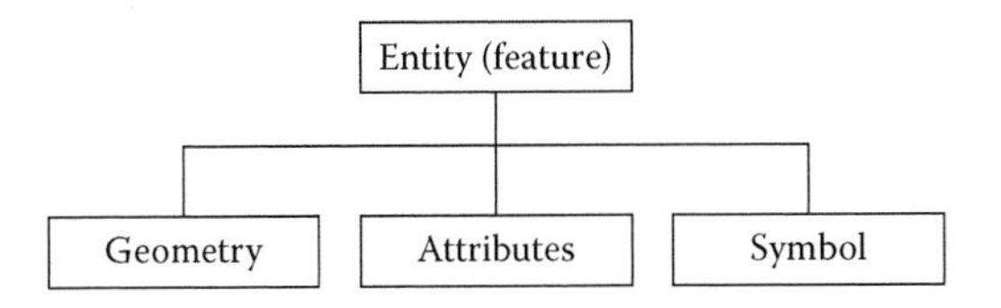

FIGURE 9.1
Structure of an entity (feature).

symbol. The entity exists as long as one of the three child elements exists. If the entity has three child elements that completely describe the geographic feature's spatial and attribute information, we consider the model to be in a healthy condition. However, if for some reason, the entity lacks a child element or its child elements cannot express real geographic features, we refer to the model as morbid. A healthy model and a morbid model can be mutually converted.

9.2.2 Child Elements of an Entity

Geometry: A geometric element is the graphical expression of GIS. In the modelling process, geometric elements express the real world through external manipulation.

Attribute: An attribute is an entity's characteristic information, such as feature names, the width of a road, or an elevation.

Symbol: A symbol is an entity's data presented as a map. During the modelling process, the initial generation of such data results from the geometric and attribute elements according to a certain rule (commonly known as the symbol library). This initial process is also known as map symbolization. However, in map presentation theory, the initial generation is not enough to represent the spatial data. Thus, the map symbolization data need additional processing to guarantee the correct relationship between each symbol. In this way, the symbols can be accurately presented. The processing operations include essential cartographic generalization, editing, and so forth.

9.2.3 Relationships between Elements

An entity element manages its child elements. It indirectly couples the relationships among child elements. The entity element and its child elements can communicate with each other, which means that the entity can receive, store, and deal with information about its geometry, attributes, and map symbol. This information is mainly related to the existence, position, and changing attributes of a child node. It is necessary for an entity to receive information on its children, as this information indicates whether the position of the symbol is reasonable and whether the map symbol and geometry are consistent. When necessary, the quality consistency should be checked according to this information.

9.2.4 Elements' Manipulation and Information Transfer

1. Manipulating an entity: In the integration model, an entity manages its child elements; thus, changes in an entity will surely affect its children. When an entity is created, its geometry, attributes, and

symbol are constructed. Similarly, editing or deleting an entity element is equivalent to editing or deleting its child elements.

2. Manipulating a child element: When a child element is manipulated, its brother elements are not affected. Geometry, attributes, and symbols are siblings; thus, there is no directly coupled relationship between brothers, which means that changing one child does not affect the other child elements. A typical manipulation is editing a child element's symbol, which does not affect the geometry and attribute elements; however, the edited information will be passed on to its parent, and the entity will store a message about the position change. This information provides a basis for later inspection and quality control. In some cases, the geometric element is deleted, but this still does not affect the existence of the other brothers (in a traditional model, the symbol relies on the geometry). We call this entity a morbid entity; however, it still exists, and the entity stores the information that the geometry has been deleted. In the following inspection and quality control, we can make the entity obtain a healthy status by adding its geometry and through appropriate editing.
3. Manipulating the information stored in an entity: An entity manages the information of the entire model. The entity retains information about the integrity and changes of its child elements. During external processing, we rationally judge the entity integrity and take certain measures according to this information.

9.2.5 Summary of Consistency Quality Control Mechanisms of DLGM

The structure of the integration model, the relationships between each element, and the information flow pattern constitute the DGLM product's consistency quality control mechanism.

9.3 Quality Control during the DLGM Process

9.3.1 DLGM Manufacturing Process

The DLGM is produced in an integrated production system. The production process is developed through a large number of operations. The flowchart is shown in Figure 9.2.

The whole progress contains three steps: (1) an integrated template is constructed based on the cartographic symbols and data dictionary; (2) the DLGM object is automatically created based on the template; and (3) the DLGM product is generated after the data are edited and the quality

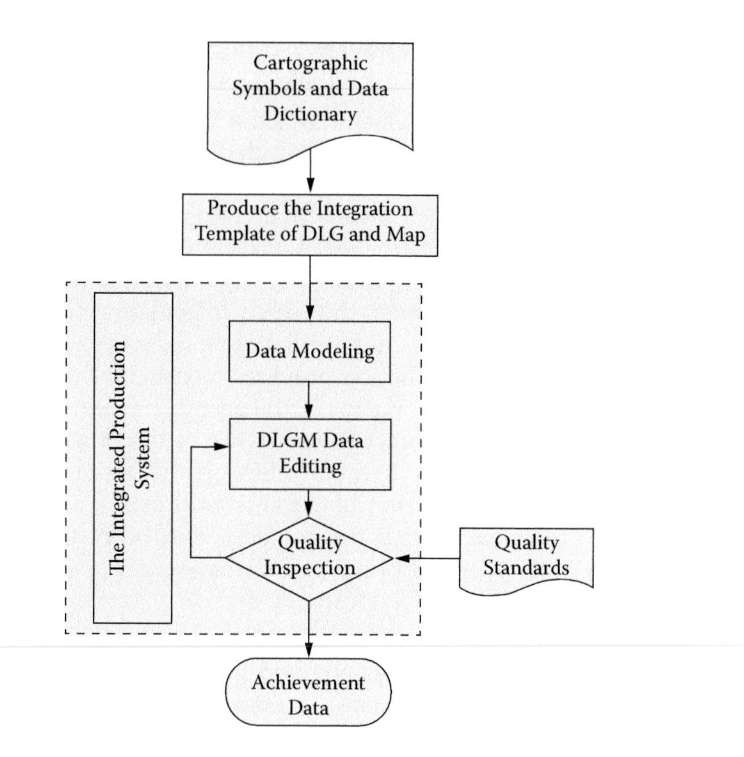

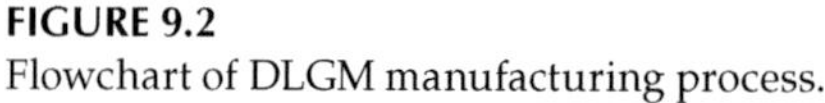

FIGURE 9.2
Flowchart of DLGM manufacturing process.

checked. Each step in the DLGM production is essentially a measure of data quality, which is important for the quality control of the DLGM. All the steps are essential for ensuring the DLGM's data quality.

9.3.2 Producing the Integrated DLGM Template

At the beginning of DLGM production, an integrated template is developed strictly according to the data dictionary and cartographic symbols. For vector data, the type of geometry, layers, attributes, and symbols are defined in the template. Map symbols are precisely defined according to cartographic symbols, including point symbols, line symbols, polygon symbols, text symbols, and their color, size, direction, and generation.

This integrated template completely expresses the content of the data dictionary and cartographic symbols. During the data modelling process, DLGM objects are created according to the prepared template. This guarantees the consistency of the DLGM. Finally, during the quality inspection and analysis, the template can be used as a standard. Integrated template quality control mainly includes the following aspects.

9.3.3 Data Modelling

DLGM modelling consists of automatically transforming raw data into DLGM objects according to the template in the integrated production system. A DLGM object is constructed from its geometry, attributes, and symbol. The symbol is created according to the geometry of the DLGM, which ensures the standards of the symbol and maintains the consistency of the DLGM.

9.3.4 Data Editing

Imperfect raw data, data changes and updates, and conflicted relationships among features lead to the necessary work of editing DLGM objects that are created by the template during DLGM production. This mainly includes the following aspects: updating the geometry and spatial topology and adding attributes, symbols, and so on. In our study, many novel processes are also designed to ensure the quality of DLGM products.

Editing the spatial relationships among the data often involves public edges, which are traditionally defined by copied features. In the integrated production system, a new process, the geometric construction process, has been developed. This process constructs complex lines based on public lines that have been captured once. A complex line is used in different layers. This method ensures the topological relationships and eliminates gaps between features, so the quality of the vector data is guaranteed. As shown in Figure 9.3, the public sides of the four layers (BOUL, HYDL, TERA, and VEGA) have been captured once. When the public side changes, the four layers change simultaneously.

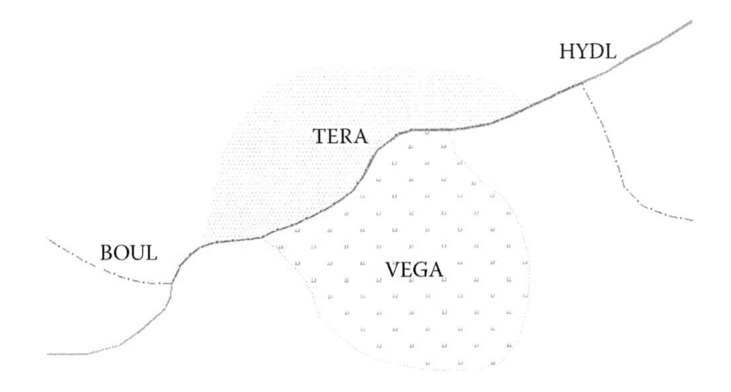

FIGURE 9.3
Public side of features.

9.3.5 Content of Quality Inspection

1. Quality inspection of integrated template.
   a. Definition of features' attributes: Check the map product to determine whether the layer features' attributes meet the quality inspection requirements.
   b. Inspection of features' symbols: Check whether the features' symbols meet the requirements for size, color, location, and annotated font.
   c. Map contour decoration: Check whether the symbols meet the requirements of the contour decoration schemes.
2. Quality inspection of DLGM products: The quality inspection of DLGM products mainly includes the DLG, map data, and integrated data.
   a. The DLG is checked by the mapping product inspection and acceptable provisions.
   b. The inspected content of the integrated map includes the consistency, correspondence, and integrity between the DLG and its corresponding map symbol. Furthermore, the consistency between the DLG and map is checked with respect to location, geometry, semantics, and attributes. The correspondence between features and symbols is checked with respect to faulty or missing symbols, or multiple correspondences of a symbol.
   c. The integrity is checked in both the features' attributes and symbols.

9.3.6 Quality Inspection

Quality inspection is an indispensable part of the data production process.[3] There are some inevitable errors in the data production process, such as inaccurate models, unreasonable geometry, and conflicted topological relationships. We have conducted a lot of research on the DLGM production process and have provided specialized quality control methods and processing tools to ensure the quality of DLGM products.

Because the information on geometry, attributes, and symbols is stored in the DLGM object, it can be used to check for DGM consistency. Moreover, methods for checking for inconsistencies between features and their corresponding symbols or labels and attributes have been developed.

In the production system, the DLGM product is inspected according to the quality control specifications. In addition to common methods for checking the quality of conventional vector data, we provide a method for checking maps.

9.4 Conclusion

A DLGM is an integrated product of GIS and vector map data. The quality of this product is an important factor in its production process. In our integration production system, the topological map model, which is the basis of DLGM products, adapts to data modelling and editing. MapStore is the production platform, and the integrated template is an important tool for guaranteeing quality. Our method provides accurate data and autopreserves the spatial relationships between data. The constant improvements in DLGM products' quality control systems ensure that the products also improve.

Acknowledgment

The research work presented in this chapter was supported by the Geographic National Condition Monitoring Engineering Research Center of Sichuan Province (GC201515).

References

1. W. Wu and X. Guo. 2011. An introduction to national fundamental geographic information digital product: Digital line graphs and map. *Bulletin of Surveying and Mapping*, pp. 61–63.
2. J.-S. Hu and J.-R. Kang. 2005. Error analyzing and quality control to map scanning digitization. *Science of Surveying and Mapping*, 30(2), 90–91.
3. W. Shi, P. Fisher, and M. F. Goodchild. 2004. *Spatial Data Quality*. Boca Raton, FL: CRC Press, pp. 85–95.

10

VGI for Land Administration: A Quality Perspective

Gerhard Navratil and Andrew U. Frank

Department of Geodesy and Geoinformation,
Vienna University of Technology, Vienna, Austria

CONTENTS

ABSTRACT The use of volunteered geographic information (VGI) or crowdsourced data (Goodchild, 2007) has become a common approach for data collection in recent years. This has caused several people to propose the use of VGI to collect data for various fields. Success stories were the (nongeographic) Wikipedia encyclopedia and OpenStreetMap (OSM). The use of VGI in land administration has also been proposed. Robin McLaren (2011) proposed crowdsourcing as a way to increase citizen collaboration for land administration to enhance transparency at lower costs. Keenja et al. (2012) discussed the perception of VGI within the Dutch cadaster. Basiouka and Potsiou (2012) even discuss how crowdsourcing can be used to identify errors in the Hellenic cadaster. One problem of VGI is quality control (compare Goodchild and Li, 2012). Only a small group of people can verify the correctness of specific types of information in a land administration system. A cadastral boundary line, for example, marks the border between two pieces of land (parcels). The correct location of the boundary can only be assessed by the owners of these parcels or by a surveyor after investigation and measurement. How can VGI provide reliable data if the correctness cannot be checked easily? Only boundaries between areas of different land use may be visible and thus verifiable, but land administration is equally dependent on ownership boundaries. In this chapter, we discuss the types of data used in land administration as defined by Dale and McLaughlin (1999). These categories are then analyzed to identify the areas where VGI can actually provide reliable input. Such an analysis shows how to use the methodology of crowdsourcing for land administration and clarifies the limits of such use.

10.1 Introduction

Volunteered geographic information (VGI) is a new phenomenon (Goodchild, 2007), enabled by the increasing capabilities of Web 2.0. There are success stories for the application of VGI, for example, the Wikipedia encyclopedia or the OpenStreetMap (OSM) mapping project. Encouraged by these examples, additional fields of application have been proposed and a number of authors have suggested the use of VGI for land administration (e.g., Basiouka and Potsiou, 2012; Keenja et al., 2012; McLaren, 2011). However, there is a fundamental difference between an encyclopedia like Wikipedia and a subsystem of public administration: errors in an encyclopedia are annoying but can be corrected quite easily when detected. Errors in public administration, however, may have dramatic consequences. The result of a decision process of an authority in public administration is a governmental act, and it has immediate legal consequences to third parties. The act of marriage is usually performed by a spoken sentence, for example, "I declare you wife and husband," accompanied by signatures on a document. The result has legal effects for the two persons, but it is also valid against third parties. VGI cannot have such an effect. VGI may be useful even if there are typos in names. A misspelled name in a land register may provide the legal basis for a fraudster to sell land he does not own; as a result, the rightful owner loses his land or the faithful buyer loses his money, even if neither of them did anything wrong, and a state guarantee may have to cover the loss. Therefore, data in public administration must be as reliable as possible and must pass strict checks; for all changes, a record of who changed the data, and for what reasons, is necessary for later audits. Goodchild and Li (2012) presented methods to check the quality of VGI. In this chapter, we discuss which of these methods are applicable to land administration, and for which parts of land administration VGI may be a viable option.

10.2 Land Administration

Land administration is an integral part of public administration. "The careful management of land and property is fundamental to economic development and the sustainability of the environment" (Dale and McLaughlin, 1999, p. 2). Land administration is related to various public activities like land tax, land market, land use planning, and the protection of rights on land. It usually consists of legal, environmental, and geometrical information. Legal information is focused on ownership and other land-related rights, including mortgages or rights of way. Environmental information covers aspects like quality of soil, pollution, land cover, and noise. Geometrical information describes the subdivision of land into smaller pieces (called parcels) as a basis to store legal information. For each parcel, at least location, size, shape, and extent are registered.

The topical separation is usually reflected by organizational separation as well. Civil law rules are applied to most legal information. Public law restrictions (e.g., development rights) are the exception, since they are granted by administrative bodies. Nonlegal information, however, is always maintained under public law regulations. Thus, in many jurisdictions, legal information is handled by courts and the remainder by one or more administrative bodies.

Legal information on rights to land has two important aspects:

1. It is always restricted to a specific area and must therefore contain a boundary. The right of ownership of one person ends where the right of ownership of his neighbor begins. It may not be clear, however, where the boundary line is located exactly.
2. The rights described are not visible in the physical world. When identifying the area owned by one person, we usually have to refer to visual demarcations like walls, fences, or differences in the use of the land. However, these demarcations may not faithfully indicate where the boundary is, and they may not even exist. Figure 10.1 shows an example for such a situation. The orthophoto on the left shows a large, homogeneous area. It could be assumed that since it seems to be used as one piece, it is owned by one person. However, the cadastral data on the right reveal that the area consists of four different parcels. Each of them may have a different owner, and one person may simply have rented them and is now using them collectively.

The geometrical information depends on the legal information. A description of a boundary line can only be created if the spatial extent of the ownership right is known. Once the boundary is described, the size of the parcel is easy to define, although different methods must be used depending on

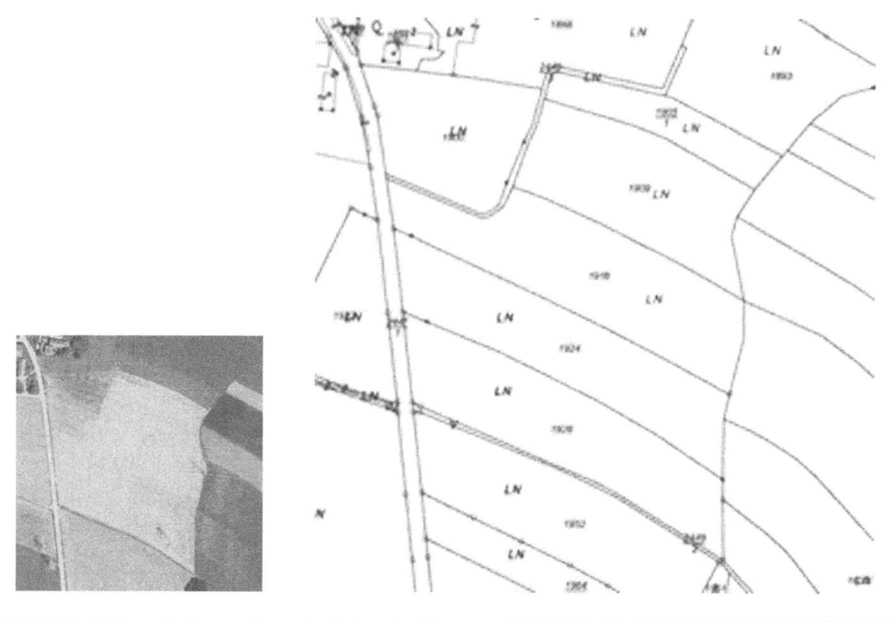

FIGURE 10.1
Observation of an area (left, orthophoto) and cadastral data of the same area (right).

the type of boundary description (compare Navratil, 2011a). Assessing the geometrical quality of cadastral data is not straightforward. Since the earth is not stable, geometries will change over time. Thus, any data set defined at a specific date will get outdated and positional accuracy will decrease. An extreme case for such a situation is the cadaster in New Zealand where the effects of plate tectonics require a dynamic reference frame (Grant et al., 2014). However, even slow changes of the reality or necessary adaptations in the representation can result in significant discrepancies between a real situation and a legal situation (Navratil et al., 2010; Navratil, 2011b).

Environmental information is necessary to use land efficiently and to define necessary restrictions for environmental sustainability. It is used in administrative processes, for example, for the following:

- Land tax: The taxation of landownership must be based on equitable and transparent rules. There are two different approaches in use: One approach assesses land tax based on an approximate market value of the parcel. This approach requires a system of mass valuation to keep the values updated. Mass valuation uses a variety of data, including sales prices, object descriptions, and data describing the nearby infrastructure (Brunauer et al., 2012). The other approach is based on the productive value of land. This is well suited for agricultural areas where the quality of the soil determines the production capability of the land.

- Land use planning: Making optimal use of land in densely populated areas is necessary in order to restrict urban sprawl and protect areas used for food production. The typical instrument for this is land use planning. The result of land use planning is a map defining the spatial distribution of processes. The creation of a residential building, for example, is restricted to residential areas, thereby protecting agricultural areas and securing food production capability. The basis for land use planning is information on existing objects and their current and potential uses.
- Agricultural subsidies: Food production is an important aspect of a national economy. With growing populations and growing cities, suitable areas for food production are becoming rare. Therefore, even unfavorable areas are used for food production. However, these areas require more effort and produce less food. In order to make them profitable, agricultural subsidies are paid to the farmers. Subsidies are calculated based on area, slope, soil conditions, climate, and so forth.

10.3 Volunteered Geographic Information

Like crowdsourced data, VGI are collected by a large number of people (Goodchild, 2007), but these people normally have no specific training for the collection process. As a result, the tools used for data collection and integration must be widely available, easy to use, error resistant, and transparent. A good example for the simplicity of the tools is the Wikipedia encyclopedia, one of the earliest successful crowdsourcing projects. Once the text for a specific entry is formulated, the majority of the work is done. Uploading the text to the encyclopedia and linking the text to other entries is done quickly. Thus, the topical knowledge of regular citizens is efficiently collected.

Most VGI initiatives concentrate on collecting observations. OSM collects the location and type of transportation infrastructure, while platforms like Flickr and Picasa provide access to georeferenced images and videos. VGI initiatives are only successful if three components work together:

- A large group of people to provide new data and to check, extend, and (if necessary) correct existing data
- Tools that allow people to perform their editing tasks with as few new skills as possible
- Methods to check the quality of the entered data

Goodchild lists a number of possible reasons why people contribute to VGI (Goodchild, 2007): self-promotion, personal satisfaction, and altruism.

Currently, the number of users of OSM, for example, is still growing rapidly and reached 2 million registered users in January 2015. With an increase in users, the percentage of users contributing (but not their actual number) decreases: from an all-time high of almost 14% of all users contributing in 2007 and 2008 to only 1% in January 2015 (OpenStreetMap Foundation, 2015). However, the reduction in the percentage of users contributing is not a cause for concern and is observed in other similar VGI systems as well. However, it has an effect on the quality checks because the majority of the users will not likely report errors in the data, and errors can only be eliminated if they are noticed.

10.4 Quality of VGI Data

The quality of VGI is a widely discussed topic. Numerous examples of quality checks for different kinds of data sets can be found in literature, and different approaches are used. Bishr and Janowicz (2010), for example, use informational trust to assess information quality. Goodchild and Li (2012) categorized the approaches. They distinguish between the following methods:

1. Crowdsourcing approach: Data may be correct if a large group of people agree on it. The principle has been successfully applied by the Wikipedia encyclopedia. It does not work equally well on geographic data collection since "it will apply best to geographic facts that are prominent" (Goodchild and Li, 2012, p. 113) and can therefore be observed and reported by many.
2. Social approach: Trusted users act as gatekeepers for information entered by other users. These trusted users can, for example, contribute heavily and thus have more experience than others. These users can easily be distinguished; Mooney and Corcoran (2012), for example, reported that 11% of the contributors were responsible for 87% of the changes on heavily edited (more than 15 times) data in OSM. Nevertheless, users that initiate many changes may have a vested interest in doing so and are therefore not automatically the most objective observers.
3. Geographic approach: The provided information is compared to the existing geographic knowledge. Motels, for example, should be close to a road and water does not flow uphill. Data that are not plausible must be flagged and investigated.

The third approach has been used, for example, by Jackson et al. (2013) to assess completeness and spatial error of VGI. A variation of this approach is the quality control by comparing VGI to official data. Quality assessment for OSM data typically uses this approach. The first analysis of OSM data quality was probably done by Haklay in 2008 (Haklay 2010). A detailed analysis on building footprints showed the complexity of the problem, ranging from scale and shape to semantic and positional accuracy (Fan et al., 2014). The analysis concludes that the quality is still lower than the quality of the official data set for the test area (the city of Munich in Germany) because buildings are simplified, semantic accuracy is below 60%, and there is an average horizontal offset of 4 m. However, determining the building footprint is a quite simple problem compared to data needed for land administration. It can be assumed that the term *building* is well understood and has no spatial variation in the geographical extent of a single city. How does VGI then work for data used in land administration, which applies concepts to a whole country? To understand the quality issues, one must ask questions such as the following:

- Can laypersons identify the object of relevance?
- Is it possible to obtain data with the required precision?
- Are the entered data trustworthy?

10.4.1 Precision of VGI Data

Modern positioning devices are cheap, simple to use, and reliable. Laypersons can use a Global Navigation Satellite System (GNSS) receiver and determine the position of boundary indicators like walls and fences. They can upload the observed data to a database if there is a suitable web interface that helps them through this process. OSM is an example of such a system. However, there are two major differences between OSM and land administration data:

- Precision of the coordinates
- Relevance of the visible objects

OSM aims at providing a map of the road network. The scale for such a map is usually at least 1:10,000, and there is a strong relation between the quality of the observation and the scale of the data (Frank, 2009). A positional uncertainty of 5 m is acceptable for OSM because in most cases the point will be located within the relevant object it represents (e.g., the road). Boundaries needed for landownership must be known with a higher precision to avoid conflicts between neighbors. In densely populated areas, uncertainty must be much better than 1 m. It is possible to achieve high precision in the range of centimeters by using GNSS devices. However, this requires sophisticated observation and analysis strategies that are difficult for untrained people to execute, as well as expensive equipment. Although providing the required training would be possible, the time investment necessary and the cost of the equipment needed for the required precision restrict the number of persons willing and able to collect boundary data. VGI would thus be restricted to topics where a highly accurate positional reference is not necessary.

10.4.2 Completeness of VGI Data

An important aspect of data in land administration is completeness. All landowners, for example, should be paying land tax. It is a problem for the tax system if the data set used to impose the tax is not complete because some landowners will have to pay tax while others do not. This aspect of completeness is known as omission. VGI is based on the willingness of citizens to provide information. No single citizen has complete knowledge. Citizens may even be reluctant to provide information if the result is negative for them (i.e., if they have to pay higher taxes). In this case, it cannot be assumed that the shared knowledge is complete.

The opposite of omission is commission, that is, excess items in the data collection. This will also happen with VGI because objects in reality may be destroyed or merged, and this must be reflected in the data set. However, the only people who directly observe the destruction are those who are part of that process. Other people would have to compare the data in the data set with reality to identify the change.

10.5 Checking Entered Data

10.5.1 VGI for Data Describing Rights

Laypersons can only document visible objects. Walls, fences, and other visible objects are indicators for (ownership) boundaries, but they do not necessarily coincide with them. Rights themselves are not observable. One can observe a person living in a house, but what follows from that for the legal status? The person may be owner of the land (including the house), have rented the house, or simply be a guest of the owner or the tenant. These are different legal situations. One must legally differentiate between right of ownership and right of use, but it is impossible to distinguish between the cases by observation of behavior only. Thus, additional knowledge must be used to separate these cases. There are two possibilities of how to get that information: the easiest way is asking the person living in the house. This has a major shortcoming: everybody who wants to check the registered information would have to ask anew. How long would it take until the person living in the house gets upset? The information obtained from the occupant is also not reliable; the person found on the premises may have an interest to pretend that they are the owner, even though they are not. If the data are marked by the reliability which the informant believes the data to have, then higher-reliability judgment comes from persons adding the data with closer connections to the occupant (family or friend), which in turn reduces the trustworthiness, given the interest of the persons involved.

Misconceptions about landownership and the process of selling and buying land may further introduce errors into the data collected by VGI. The occupant may, for example, have concluded a sales contract with the previous owner and therefore believes that he is the legal owner. In most legal systems, transfer of ownership of land, unlike transfer of ownership of other goods, requires a written contract with additional qualifications. Often it must be concluded before an authority and later registered; only then is the transfer from seller to buyer complete. In many countries, private sales—not in the proper form and not registered—are common, and all parties believe that the buyer is now the owner because he bought the land, but legally the seller is still the owner of the land because the contract is not valid or not registered, and therefore transfer of ownership has not in fact occurred.

The classification of rights is difficult, but the spatial delimitation of these rights is even more difficult. From the observation that a person uses a specific piece of land, it can neither be concluded that the person is entitled to use the whole area nor be concluded that this is all of the area the person can use.

10.5.2 VGI on Physical Objects and Their Properties

Observation of physical objects is direct. A fence, a wall, or a building is usually visible at a distance of tens of meters. However, laypersons can only take pictures of these objects. If they want to observe the exact position of these objects, they would have to use technology like GNSS. The application of such technology requires the physical presence of the person at the location of measurement. This is not a problem for a fence separating private property from public space, but what should be done with a fence separating two private properties? In many jurisdictions, surveyors are permitted to walk on privately owned land if this is required for their work, but laypeople producing VGI do not have this right and may get sued for trespassing. Therefore, only neighbors or other people with permission to enter the property can measure the position of a fence. However, this is not an independent check.

One solution is to provide orthophotos. Their creation can be organized by public bodies, and the VGI contribution is the digitizing and modelling of physical objects (compare Konecny, 2012). However, this method cannot guarantee that the objects registered are used to delimit a right. Fences may be placed inside the land parcel for practical reasons, for example, for driving without the need to open gates. As shown in Figure 10.1, there may even be numerous boundary lines that are not observed because they are not visible.

Another method could be evaluation of amateur photographs. Objects that are visible from several angles can be reconstructed at least semiautomatically.

The advantage of this method would be that the huge database of already existing, georeferenced photographs can be used. Erroneously georeferenced photographs may be a problem, though; photographs that show completely wrong places will typically have no others to form a group that can be evaluated. However, if one image is slightly shifted, it will have an influence on the positional quality of the result. This problem may be solvable when combining several objects and a detailed investigation is warranted.

Land cover is easy to observe. Building construction is visible from a distance, and growing trees, vines, and so forth, are similarly observable. This information can even be extracted from images. More difficult is the observation of land use, which is related to but not the same as land cover. Land use describes activities occurring on the land and how it is classified. When does a building become "commercial use"? When is a piece of land stocked with trees called a "forest" (and under which legal definition)?

The advantage of VGI for land use and land cover data is the implied classification from a visible situation to a land use or land cover class. People familiar with the general situation may provide good judgment, provided they understand the (legal) definitions for the land cover and land use classes. It is likely that some training is required to obtain reliable, repeatable results.

10.5.3 VGI on the Environment

Environmental information can also be collected by VGI. At least in theory, laypersons can collect soil or water samples for quality testing. Fraud can be detected if enough samples are collected. However, the problem of trespassing exists in this case as well. Taking soil or water samples requires a physical presence that may be prohibited by the landowner. Samples must be analyzed, which usually requires some equipment and training.

The quality of results can be checked by relying on spatial autocorrelation. A specific soil type on a parcel suggests that the neighboring parcels have similar soil types. This can be used for automated plausibility checks of the data. Given that the equipment and skill requirement for testing is high, it is unlikely that the group of contributors is large and that many overlapping results are collected.

10.6 Practical Issues

Quality is an important consideration for information in general and geographical information in particular. This problem is not automatically solved by using VGI. The Wikipedia encyclopedia, as one of the most successful examples of crowdsourced data, sometimes faces major edits (so-called edit wars) inserted by followers and opponents of politically active persons

(Garber, 2012) and has finally developed specific procedures to achieve equitable results.

How would VGI for land administration data deal with the interests of parties? We have already mentioned that occupants may try to enter data showing them to be owners (and sometimes owners pretend to be mere renters to avoid property taxes). More subtle are cases where a business opportunity is closely related to data stored in a land administration database. It may be tempting for competitors to create opportunities for themselves or to prevent opportunities for their competitors by falsifying land administration data. Although false data may eventually be corrected, the delay may be sufficient to create a business advantage for one party. What does this mean for a public administration's liability? Should liability charges be pressed against the original contributor? A solution is necessary, but it may have an impact on the willingness of citizens to contribute data, if legal action against a person contributing VGI is possible.

To approach this problem, one must consider the principles of public administration and due process. A decision by a judge (in a civil law case) or an administrator must follow the law. Judges and administrators are required to investigate the facts of the case. If the data provided by VGI are incorrect, correction is required before a decision is made. Two issues have to be analyzed: whether the information in the administrative system was wrong and when the correction was made.

Assume the official registry does not correspond to the information provided by VGI. The VGI then provides:

- Correct information: The information provided by VGI corrects wrong information or adds facts previously unknown to the authority.
  - The authority may be obliged to perform an on-site inspection before deciding on the application. In this case, VGI has no negative effects, but it prompts the authorities to do a spot-check (which may be perceived by them as a nuisance).
  - The authority may also choose to decide with the available (wrong or incomplete) data only. In this case, incomplete VGI, which is available for some areas and not for others, may result in decisions that are not equitable. Assume, for example, that someone reports that a bird species protected by a wildlife conservation act is breeding at a specific location. The person may have missed that birds of the same species breed in neighboring locations as well—thus the VGI is incomplete, which is part of the character of VGI. In such a case, the available information affects a decision for one parcel but not for others. In principle, the administration should investigate and ensure that the data cover all the area, but the likelihood of an investigation to confirm the absence of a bird species is low.

- Incorrect information: The information provided by VGI contradicts data found in the public registry, but the data in the registry are correct and the VGI is wrong. This may be done by accident or on purpose, for example, by willful deception or intentional misleading. Thus, it would be actionable, although it may be difficult to prove that it was done intentionally. Accidentally providing wrong information should not be pursued because this would contradict the idea of VGI.

Since currently available methods for quality control of VGI may fail due to the specific nature of land administration data, one might rely on the contributors' levels of trustworthiness. Users who have contributed valuable and correct information in the past are trusted. However, instead of acting as gatekeepers for others (which is denoted as the social approach for quality control by Goodchild and Li [2012]), only their VGI would be entered in public registries. To satisfy the requirements of audits, only data from contributors for which full personal details are known could be entered in public registries. Knowing that one's actions and the information entered are kept on record reduces the incentives to enter misinformation for personal gain.

Finally, one must consider the effects of VGI on the process of decision making. Adding VGI in the course of a decision process has potentially two effects: (1) Regardless of whether the data contributed are correct or wrong, new data may prompt the decision maker to investigate, which certainly delays the decision. (2) An applicant for a building permit, for example, may have investigated public records and found that his application—prepared at some cost—is likely to be granted. VGI contributed during the decision process may uncover additional facts that would lead to his application not being granted in the end. The legal question is whether his "good faith" should be protected and the decision made with the data available when the application was filed or whether the material correctness of the decision is more important.

Public authorities must hold up to audits and must demonstrate where the data they used came from and how they used the data. It was already mentioned that only data from sources that are fully known can be used in administrative decisions. However, asking contributors to disclose their individual identities goes against some of the spirit of VGI. The requirements of auditors lead to database designs where the temporal sequence of additions to the database can be reconstructed; it must be possible to determine which facts were known when a decision was made. Technically, this is achieved either by storing all updates or by storing a snapshot of the relevant parts of the database at the time of decision. The first approach is used by numerous systems, including Wikipedia or cadastral systems.

10.7 Conclusions

VGI approaches can be used for the collection of some types of data; the question explored here is whether the specific type of data used in land administration is suitable for VGI. Volunteers can only provide information they have or can easily obtain from general information derived from observation or communication. Thus, invisible facts, for example, regarding ownership, can only be provided by very few persons, who are typically close to the situation and have substantial vested interests.

The nature of land administration data provides problems for quality control when data are volunteered by persons outside of the administration. Land administration is based on a variety of data:

- Rights and their spatial extent
  - Ownership
  - Usage rights
  - Mortgages
- Physical objects and their properties
  - Buildings
  - Land use
- Environmental information
  - Soil
  - Water
  - Noise

Some of these data are used in legal (civil) proceedings for landownership and other private rights, taxation and land use planning, and the determination of land use subsidies for agriculture. The applications of land administration data and services started with land tax and property registration and today include support of the land market, environmental protection, and agricultural subsidies (compare Muggenhuber et al., 2011). Thus, the above list can only be a snapshot of data currently used, and the list may be extended in the future. This is a challenge for public data collection, and a (cost-free) contribution from VGI is welcomed by the authorities.

VGI can provide information on topics where direct observation is sufficient but is difficult for facts that are not directly visible. Land cover and land use, for example, can be observed, and thus VGI processes can collect the information. It may be difficult to verify the data using VGI with a "crowd sourcing approach" (Goodchild and Li, 2012) since there may only be a limited number of citizens able to make the observations. The result could be data sets complementing information from public sources. These additional data sets can provide a different perspective, for example, land use data in contrast to land cover data provided by remote sensing.

The discussion showed that it is not possible to base land administration completely on VGI. Rights on land, which constitute a large portion of data used in land administration, cannot be observed directly by citizens with local knowledge because they are not observable. Thus, the quality of VGI is impossible to assess, for example, for land taxation. VGI can support traditional data collection mechanisms in situations where observation is possible but heavily time dependent. However, it is necessary to take precaution against (intentional or accidental) falsification of data in the database. To make a decision, the authorities must verify the data; the role of VGI in land administration is therefore mostly one of giving notice of incomplete or wrong data in public records, with safeguards to prevent malicious uses of this new information channel to land management authorities.

References

Basiouka, S., and Potsiou, C. 2012. VGI in Cadastre: A Greek experiment to investigate the potential of crowd sourcing techniques in cadastral mapping. *Survey Review*, 44(325), 153–161.

Bishr, M., and Jankowicz, K. 2010. Can we trust information? The case of volunteered geographic information. In A. Devaraju, A. Llaves, P. Maué, and C. Keßler (eds.), *Towards Digital Earth: Search, Discover and Share Geospatial Data, Workshop at Future Internet Symposium, Berlin, Germany, CEUR Workshop Proceedings*, vol. 640, paper 3.

Brunauer, W., Feilmayr, W., and Wagner, K. 2012. A new residential property price index for Austria. *Statistiken, Daten and Analysen*, 12(Q3), 90–102.

Dale, P., and McLaughlin, J. 1999. *Land Administration*. Oxford University Press, Oxford.

Fan, H., Zipf, A., Fu, Q., and Neis, P. 2014. Quality assessment for building footprint data on OpenStreetMap. *International Journal of Geographical Information Science*, 28(4), 700–719.

Frank, A. U. 2009. Scale is introduced by observation processes. In R. Devillers and M. Goodchild (eds.), *Proceedings of the 6th International Symposium on Spatial Data Quality (ISSDQ)*. CRC Press, St. John's, Newfoundland and Labrador, pp. 17–29.

Garber, M. 2012. Paul Ryan, "Brown noser"? The Wikipedia edit wars begin for Romney's running mate. *The Atlantic*, August 11.

Goodchild, M. F. 2007. Citizens as sensors: The world of volunteered geography. *GeoJournal*, 69(4), 211–221.

Goodchild, M. F., and Li, L. 2012. Assuring the quality of volunteered geographic information. *Spatial Statistics*, 1(1), 110–120.

Grant, D., Crook, C., and Donelly, N. 2014. Managing the dynamics of the New Zealand spatial cadastre. In S. Winter and C. Rizos (eds.), *Proceedings of Research at Locate '14, Canberra, Australia, CEUR Workshop Proceedings*, vol. 1142, pp. 60–71.

Haklay, M. 2010. How good is volunteered geographical information? A comparative study of OpenStreetMap and Ordnance Survey datasets. *Environment and Planning B: Planning and Design*, 37(4), 682–703.

Jackson, S. P., Mullen, W., Agouris, P., Crooks, A., Croitoru, A., and Stefanidis, A. 2013. Assessing completeness and spatial error of features in volunteered geographic information. *ISPRS International Journal of Geo-Information*, 2(2), 507–530.

Keenja, E., De Vries, W., Bennet, R., and Laarakker, P. 2012. Crowd sourcing for land administration: Perceptions within Netherlands Kadaster. *FIG Working Week 2012*, Rome, TS03B.

Konecny, G. 2012. Alternatives for economic boundary determination in the establishment of a cadastral system. *FIG Working Week 2012*, FIG, Rome, TS07E.

McLaren, R. 2011. Crowdsourcing support of land administration: A partnership approach. *VGI (Österr. Zeitschrift für Vermessung und Geoinformation)*, 99, 1–12.

Mooney, P., and Corcoran, P. 2012. Characteristics of heavily edited objects in OpenStreetMap. *Future Internet*, 4(4), 285–305.

Muggenhuber, G., Navratil, G., Twaroch, C., and Mansberger, R. 2011. Development and potentials for improvements of the Austrian land administration system. *FIG Working Week 2011*, Marrakech, TS07A.

Navratil, G. 2011a. Cadastral boundaries: Benefits of complexity. *URISA*, 23(1), 19–27.

Navratil, G. 2011b. Quality assessment for cadastral geometry. In C. Fonte, L. Goncalves, and G. Goncalves (eds.), Coimbra, Portugal, *Proceedings of the 7th International Symposium on Spatial Data Quality*, pp. 115–120.

Navratil, G., Hafner, J., and Jilin, D. 2010. Accuracy determination for the Austrian digital cadastral map (DKM). In D. Medak, B. Pribicevic, and J. Delak (eds.), *Proceedings of the Fourth Croation Congress on Cadastre*, Zagreb, Croatia, pp. 171–181.

OpenStreetMap Foundation. 2015. Stats: OpenStreetMap wiki. http://wiki.openstreetmap.org/wiki/Stats (accessed March 31, 2015).

11

Qualitative and Quantitative Comparative Analysis of the Relationship between Sampling Density and DEM Error by Bilinear and Bicubic Interpolation Methods

Wenzhong Shi,[1] Bin Wang,[1] and Eryong Liu[2]

[1]*Department of Land Surveying and Geo-Informatics, Hong Kong Polytechnic University, Hung Hom, Hong Kong*

[2]*School of Resources and Environment, Fujian Agriculture and Forestry University, Fuzhou, Fujian, China*

CONTENTS

ABSTRACT The accuracy of a digital elevation model (DEM) is at stake, critically affecting the success of DEM applications. One important issue of DEM research concerns those factors that touch the generated DEM accuracy. The relationship between DEM error and the sampling density is still a worthy question, especially for nonlinear interpolations. This research comparatively analyzes the qualitative and quantitative relationship between the sampling density and DEM error by both bilinear and bicubic interpolation methods. First, the qualitative relationships between the DEM error and the sampling density for both bilinear interpolation and bicubic interpolation models are investigated based on convergence analysis. Second, the quantitative relationships between the DEM error and the sampling density are further derived by means of the numerical approaches. Here the model error is specified by its right upper bound of the truncated error function, with the sampling density as its variable. Third, experimental studies, involving both mathematical and real DEM surfaces, are conducted to verify the foregoing theoretical findings. The theoretical derivations and experimental studies both demonstrate that the DEM quality by a bicubic interpolation method, in terms of model error, is superior to the counterpart generated by a bilinear interpolation method under the assumption of the original sample data being error-free. The new findings about the model errors for bicubic interpolated DEM, together with the previous work on bilinear interpolated DEM drawn from an earlier study, form a full picture depicting the model errors of an interpolated DEM surface. These results can serve as a guideline for interpolation model selection regarding the practical DEM production.

11.1 Introduction

The digital elevation model (DEM) is a digital representation of terrain variation where the elevation values are normally produced by interpolation based on the collected sample points. The terrain details of the earth or other planets are of great interest to researchers or engineers from various fields, including natural resources exploration, geography, geology, land management, and human health study, and even regional economic planning. Therefore, a DEM, especially a high-accuracy DEM, is currently more momentous than ever before.

In order to obtain the DEM with much higher accuracy, any discovered errors should be cut down as much as possible during the DEM construction process. Basically, the overall DEM error can be classified into the following two categories (Shi et al., 2005): (1) propagated error and (2) model error. The propagated error is the effect of uncertainties derived from the raw data or original sample data due to measurement limitations (e.g., instrument precision), while the model error depicts the differences between the artificially selected models' interpolated data and the original sampled elevation values.

Many studies have been reported on the modelling of DEM propagated errors. Li et al. (2010) gave an estimation of the mean error of a DEM that is linearly constructed from a regular grid network. Shi et al. (2005) provided a model for estimating the propagated DEM error generated by high-order interpolation algorithms. Zhu et al. (2005) presented an estimation of the average DEM accuracy, generated by linear interpolation, allowing for random errors at the nodes of the triangular irregular network (TIN) model. Kyriakidis and Goodchild (2006) further derived average line accuracy for TIN and rectangle models based on linear interpolation functions, where the errors of the component nodes are interdependent. Oksanen and Sarjakoski (2006) characterized statistical and spatial details of the errors in a fine-scale DEM derived by contour data. Carlisle (2005) discussed the limitations of using a single root mean square error to represent the uncertainty associated with a DEM and proposed a spatially distributed indicator, an accuracy surface, for DEM quality description.

With the advent of technology, more and more new measuring techniques, such as aero and terrestrial laser scanning technologies, have been developed, enabling easier access to terrain elevation data with high quality for DEM generation, whereas the previously captured DEM data were largely confined to old equipment. In this connection, the model error in this chapter is deemed as the dominant error source, or more precisely, the original collected data are assumed to be error-free.

In the field of DEM error study, Gao (1997) examined the impact of DEM resolution upon the accuracy of terrain representation and the gradients determined. The results are useful for determining the appropriate prerequisite DEM resolution for the accuracy requirements of a particular application. Huang (2000) described three methods of evaluating information loss. Hereinto, information loss is the sum of the interpolation error due to sampling a continuous terrain surface with a finite grid interval. Aguilar et al. (2006) presented a theoretical-empirical model for modelling the accuracy of a grid-based DEM. Here, the assumption was made that DEM errors were randomly distributed. However, many studies have found that DEM errors actually correlate with terrain morphology and sampling density (Hu et al., 2009; Liu and Sherba, 2012). Shi et al. (2014) investigated the accuracy analysis of the DEM with respect to its two influencing factors using an analytical approach, establishing the relationship between the interpolated DEM accuracy and spatial resolution and terrain slope. By using three theoretical approaches, Aguilar et al. (2007) estimated the 95th percentile degree of correctness, which emphasizes mathematical statistics analysis. Besides, Hu et al. (2009) introduced approximation theory from computational science to examine the model errors in DEMs interpolated by the linear polynomial interpolations upon three basic spatial objects: one-dimensional line segment (1D), triangular irregular network (TIN), and rectangle. Guo et al.

(2010) discussed the effects of topographic variability and LIDAR sampling density on DEM accuracy with respect to different interpolation methods and spatial resolutions. An adaptive interpolation strategy (Arun, 2013) has been proposed, which takes the terrain features into account for improving the estimation accuracy.

Based on the above analysis of the previous studies, the following two research issues can be identified. First, there are fewer reported theoretical analyses on DEM errors based on nonlinear interpolations for the spatial objects, for instance, (1) 1D, (2) TIN, and (3) rectangle. Second, no rigorous mathematical derivation exists relating to the relationship between interpolated DEM errors with its sampling density for either linear or nonlinear DEM interpolation methods, although an intuitive feeling has always been accepted by many researchers based on experience for a long time that DEM accuracy should be associated with its sampling intervals.

In this chapter, a systematic study of DEM error for 1D, TIN, and rectangle has been conducted based on the piecewise bicubic interpolation and approximation theory. The rest of this chapter is structured as follows. In Section 11.2, the linear and cubic polynomial interpolation methods for spatial objects, including the 1D, TIN, and rectangle, are briefly introduced, and a study of the convergence analysis of both bilinear and bicubic interpolation methods is given, correspondingly. Section 11.3 is devoted to the theoretical derivations of bicubic interpolated DEM errors based on 1D, TIN, and rectangles. After that, experimental studies described in Section 11.4 verify the foregoing theoretical findings. Conclusions are drawn and described in Section 11.5.

11.2 Interpolation Methods and Convergence Analyses

This chapter is mainly devoted to exploring the relationship between the model error (also called the truncated error) and the sampling density by the bicubic interpolation method for three spatial objects: the 1D, TIN, and rectangle. Here, a truncated error is the difference between the true elevation of a terrain surface and the interpolated surface based on the original source data.

Before proceeding, the following assumptions and notes are made for this study:

1. The original data are assumed to be error-free. Therefore, the propagated error from the original sampling data is ignored.
2. The function representing the terrain surface is at least continuous without further specification.

3. The first partial derivatives of the original elevation nodes can be provided for bicubic interpolation.
4. The analysis is only carried out on one patch. Obviously, the findings can be extended directly to all patches, namely, the entire terrain.

The modulus of continuity is first introduced for the convenience of the following proof of the relationship between interpolated model error and sampling density.

Formally, a modulus of continuity is any real-extended value function $\omega:[0,\infty] \rightarrow [0,\infty]$ used to measure quantitatively the uniform continuity of a function, vanishing at 0 and continuous at 0, that is,

$$\lim_{t\rightarrow 0} \omega(t) = \omega(0) = 0.$$

Specifically, for two arbitrary points $x', x'' \in [a,b]$, $|f(x') - f(x'')| \leq \omega(h)$. If only $|x' - x''| \leq h$, where $h = |b - a|$, then $\omega(h)$ is called the modulus of continuity for $f(x)$ on $[a,b]$. A function turns out to be uniformly continuous if and only if it admits a modulus of continuity. Therefore, $\lim_{h\rightarrow 0} \omega(h) = 0$ will hold, as long as $f(x)$ is continuous on $[a,b]$.

Interpolation methods widely used in generating digital elevation models can generally be classified as linear or nonlinear. The bilinear interpolation and bicubic interpolation methods are, respectively, linear and nonlinear methods that play important roles in DEM construction. In this section, the linear and cubic polynomial interpolation methods for the spatial objects are first described, including the 1D, TIN, and rectangle methods. For simplicity, the piecewise cubic Hemite interpolation (cubic interpolation) method is used as the representative of nonlinear interpolation.

11.2.1 Bilinear Interpolation

11.2.1.1 Linear Interpolation for Line Segment

As illustrated in Figure 11.1, suppose A and B are two given nodes; their independent variable coordinates are x_A and x_B, and their corresponding elevation values are h_A and h_B, respectively, where x_A and x_B form a one-dimensional line segment. The linear interpolation function based on the two nodes A and B is

$$l_1(x) = \frac{x_B - x}{x_B - x_A} h_A + \frac{x - x_A}{x_B - x_A} h_B. \tag{11.1}$$

Denote

$$\omega_A = \frac{x_B - x}{x_B - x_A},$$

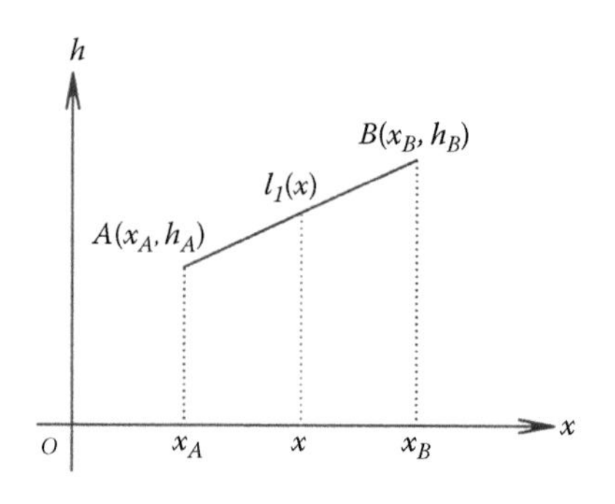

FIGURE 11.1
A linear interpolation for 1D based on two nodes A and B.

$$\omega_B = \frac{x - x_A}{x_B - x_A};$$

then $\omega_A + \omega_B = 1$.

Equation 11.1 can thus be rewritten as

$$l_1(x) = \omega_A h_A + \omega_B h_B, \tag{11.2}$$

where $\omega_A + \omega_B = 1$, $\omega_A \geq 0$, $\omega_B \geq 0$.

Suppose that $L_1(x)$ is the true interpolation function $l_1(x)$ defined on the interval $[x_A, x_B]$, and assume $L_1(x)$ is continuous within the interval, that is, $L_1(x) \in C_{[x_A, x_B]}$. For any $x \in [x_A, x_B]$, set $h = x_B - x_A$; then we have

$$\begin{aligned}
|L_1(x) - l_1(x)| &= |(\omega_A + \omega_B) L_1(x) - (\omega_A h_A + \omega_B h_B)| \\
&= |\omega_A (L_1(x) - h_A) + \omega_B (L_1(x) - h_B)| \\
&\leq |\omega_A (L_1(x) - h_A)| + |\omega_B (L_1(x) - h_B)| \\
&\leq (\omega_A + \omega_B)\omega(h) = \omega(h).
\end{aligned}$$

According to the definition of $\omega(h)$, it can be easily deduced that $\lim_{h \to 0} l_1(x) = L_1(x)$; namely, $l_1(x)$ will uniformly converge to $L_1(x)$ as h approaches the vanishing point. That is, the error, which is defined as the difference between the true elevation value and its interpolated value, is approaching zero when the sample density approaches zero.

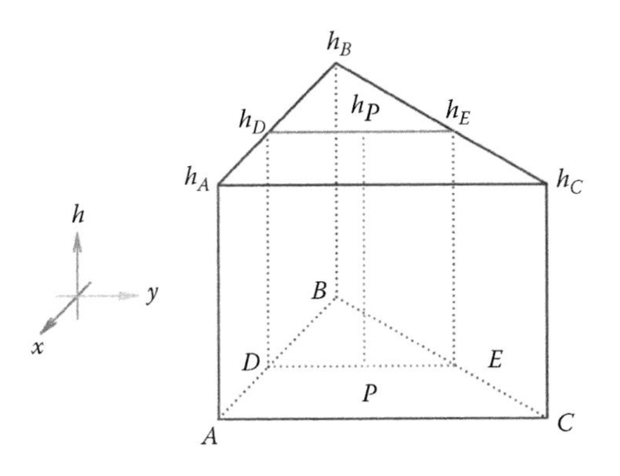

FIGURE 11.2
Linear interpolation for TIN.

11.2.1.2 Linear Interpolation for TIN

The linear interpolation for TIN shown in Figure 11.2 is also a linear combination of the elevations of the three nodes of ΔABC. Consider that A, B, and C are three given nodes of a TIN model; the independent variable coordinates (x_A,y_A), (x_B,y_B), and (x_C,y_C) constitute a triangle ΔABC, and the corresponding elevation values of its endpoints are h_A, h_B, and h_C. The linear interpolation function based on this TIN model is

$$l_T(x,y)=\omega_A h_A+\omega_B h_B+\omega_C h_C. \tag{11.3}$$

Hereinto

$$N_A=x(y_B-y_C)-y(x_B-x_C)+x_B y_C-x_C y_B,$$

$$N_B=x(y_C-y_A)-y(x_C-x_A)+x_C y_A-x_A y_C,$$

$$N_C=x(y_A-y_B)-y(x_A-x_B)+x_A y_B-x_B y_A,$$

$$D=x_A(y_B-y_C)+x_B(y_C-y_A)+x_C(y_A-y_B),$$

$$\omega_A=\frac{N_A}{D},\ \omega_B=\frac{N_B}{D},\ \omega_C=\frac{N_C}{D}.$$

It could be easily verified that ω_A, ω_B, $\omega_C \in [0,1]$ and $\omega_A+\omega_B+\omega_C=1$.

Suppose that $L_T(x)$ is the true interpolation function $l_T(x)$ defined on the TIN model, assume $L_T(x) \in C_{\Delta ABC}$ (which means that $L_T(x)$ is continuous upon the ΔABC) for any $(x,y) \in \Delta ABC$, and let $h = \max\{|\overline{AB}|, |\overline{BC}|, |\overline{AC}|\}$; then

$$\begin{aligned}
|L_T(x) - l_T(x)| &= |(\omega_A + \omega_B + \omega_C)L_T(x) - (\omega_A h_A + \omega_B h_B + \omega_C h_C)| \\
&= |\omega_A(L_T(x) - h_A) + \omega_B(L_T(x) - h_B) + \omega_C(L_T(x) - h_C)| \\
&\le |\omega_A(L_T(x) - h_A)| + |\omega_B(L_T(x) - h_B)| + |\omega_C(L_T(x) - h_C)| \\
&\le (\omega_A + \omega_B + \omega_C)\omega(h) = \omega(h).
\end{aligned}$$

According to the definition of $\omega(h)$, it can easily be deduced that $\lim_{h \to 0} l_T(x) = L_T(x)$; namely, $l_T(x)$ will uniformly converge to $L_T(x)$ as h approaches the vanishing point. That is, the error, which is defined as the difference between the true elevation value and its interpolated value of a point of TIN, approaches zero when the sample density approaches zero.

11.2.1.3 Bilinear Interpolation for Rectangle

As illustrated in Figure 11.3, the bilinear interpolation is analogous to the linear interpolation discussed above. A weighted average of the four surrounding grid points is used to determine the interpolated value. Consider that *A*, *B*, *C*, and *D* are the four given nodes of a rectangle model, the independent variable coordinates (x_1,y_1), (x_2,y_1), (x_2,y_2), and (x_1,y_2) constitute the rectangle *ABCD*, and the corresponding elevation values of its endpoints are h_A, h_B, h_C, and h_D; then the bilinear interpolation function based on this rectangle model is as follows.

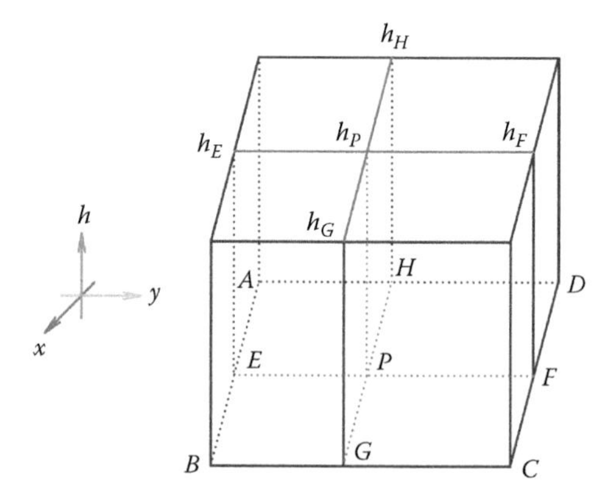

FIGURE 11.3
Linear interpolation for rectangle.

For convenience, make use of the four coefficients

$$\omega_A = \frac{(x_2 - x)(y_2 - y)}{(x_1 - x_2)(y_1 - y_2)},$$

$$\omega_B = \frac{(x - x_1)(y_2 - y)}{(x_1 - x_2)(y_1 - y_2)},$$

$$\omega_C = \frac{(x - x_1)(y - y_1)}{(x_1 - x_2)(y_1 - y_2)},$$

$$\omega_D = \frac{(x_2 - x)(y - y_1)}{(x_1 - x_2)(y_1 - y_2)}.$$

The bilinear interpolation function can be obtained:

$$l_R(x) = \omega_A h_A + \omega_B h_B + \omega_C h_C + \omega_D h_D. \tag{11.4}$$

It could also be verified that ω_A, ω_B, ω_C, $\omega_D \in [0,1]$ and $\omega_A + \omega_B + \omega_C + \omega_D = 1$.

Suppose that $L_R(x)$ is the true function of the interpolation function $l_R(x)$ defined on the rectangle, and assume $L_R(x) \in C_{(ABCD)}$. For any $(x,y) \in (ABCD)$, let $h = \max\{|\overrightarrow{AB}|, |\overrightarrow{BC}|\}$; then we have

$$\begin{aligned}
|L_R(x) - l_R(x)| &= |(\omega_A + \omega_B + \omega_C + \omega_D)L_R(x) - (\omega_A h_A + \omega_B h_B + \omega_C h_C + \omega_D h_D)| \\
&= |\omega_A(L_R(x) - h_A) + \omega_B(L_R(x) - h_B) + \omega_C(L_R(x) - h_C) + \omega_D(L_R(x) - h_D)| \\
&\leq |\omega_A(L_R(x) - h_A)| + |\omega_B(L_R(x) - h_B)| + |\omega_C(L_R(x) - h_C)| + |\omega_D(L_R(x) - h_D)| \\
&\leq (\omega_A + \omega_B + \omega_C + \omega_D)\omega(h) = \omega(h).
\end{aligned}$$

According to the definition of $\omega(h)$, we can easily deduce $\lim_{h \to 0} l_R(x) = L_R(x)$; namely, $l_R(x)$ will uniformly converge to $L_R(x)$ as h approaches to vanish. That is, the model error, which is defined by the difference between the true elevation value and its interpolated value of a point on the rectangle, is approaching zero with the sample density approaching zero.

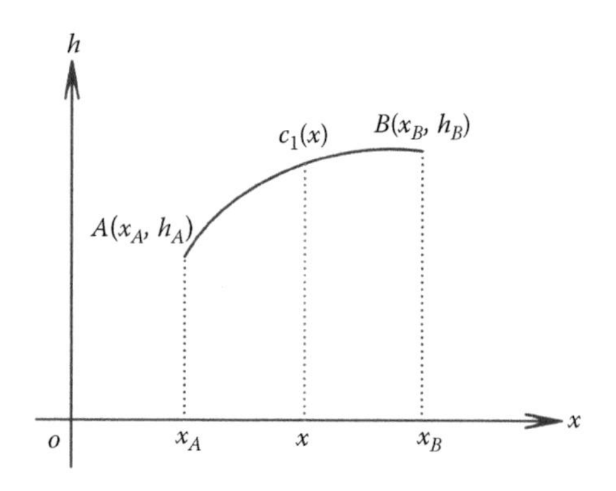

FIGURE 11.4
A cubic interpolation based on the two nodes.

11.2.2 Bicubic Interpolation

11.2.2.1 Cubic Interpolation for 1D

As illustrated in Figure 11.4, cubic interpolation for 1D, there is a piecewise cubic polynomial interpolation in that each patch is a line segment. Suppose that A and B are two given nodes of a line segment, four independent variable coordinates of them are X_A and X_B, and their corresponding true elevation values and first-order derivatives are h_A, h_B and h'_A, h'_B, respectively. The cubic interpolation function based on these two nodes is

$$c_1(x) = \left(1 - 2\frac{x - x_A}{x_A - x_B}\right)\left(\frac{x - x_B}{x_A - x_B}\right)^2 h_A + \left(1 - 2\frac{x - x_B}{x_B - x_A}\right)\left(\frac{x - x_A}{x_B - x_A}\right)^2 h_B$$

$$+ (x - x_A)\left(\frac{x - x_B}{x_A - x_B}\right)^2 h'_A + (x - x_B)\left(\frac{x - x_A}{x_B - x_A}\right)^2 h'_B .$$

For convenience, the above equation is rewritten as

$$c_1(x) = \alpha_1 h_A + \alpha_2 h_B + \beta_1 h'_A + \beta_2 h'_B , \tag{11.5}$$

where

$$\alpha_1 = \left(1 + 2\frac{x - x_A}{x_B - x_A}\right)\left(\frac{x_B - x}{x_B - x_A}\right)^2 ,$$

and

$$\alpha_2 = \left(1 + 2\frac{x_B - x}{x_B - x_A}\right)\left(\frac{x - x_A}{x_B - x_A}\right)^2,$$

$$\beta_1 = (x - x_A)\left(\frac{x_B - x}{x_B - x_A}\right)^2,$$

$$\beta_2 = (x_B - x)\left(\frac{x - x_A}{x_B - x_A}\right)^2.$$

In the following, several properties of the above coefficients are first presented, so as to expedite subsequent derivations. According to the generalized mean value inequality $(a + b + c)^3 \geq 27\, abc$,

$$\begin{aligned} 0 \leq |\alpha_1| &= \left(1 + 2\frac{x - x_A}{x_B - x_A}\right)\left(\frac{x_B - x}{x_B - x_A}\right)^2 = \frac{1}{(x_B - x_A)^3}(2x + x_B - 3x_A)(x_B - x)^2 \\ &\leq \frac{1}{(x_B - x_A)^3}\frac{(3x_B - 3x_A)^3}{27} = 1; \end{aligned} \tag{11.6}$$

$$\begin{aligned} |\beta_1| &= (x - x_A)\left(\frac{x_B - x}{x_B - x_A}\right)^2 = \frac{4}{(x_B - x_A)^2}(x - x_A)\left(\frac{x_B - x}{2}\right)^2 \\ &\leq \frac{4}{(x_B - x_A)^2}\frac{(x_B - x_A)^3}{27} = \frac{4}{27}(x_B - x_A) = \frac{4}{27}h \end{aligned} \tag{11.7}$$

It is also easy to deduce that α_2 and β_2 have the same properties as above. What is more, $\alpha_1 + \alpha_2 = 1$.

Suppose that $C_1(x)$ is the true function of $c_1(x)$ defined over the interval $[x_A, x_B]$, and assume $C_1(x) \in C_{[x_A, x_B]}$. For any $x \in [x_A, x_B]$, let $h = x_B - x_A$; then we have

$$\begin{aligned} |C_1(x) - c_1(x)| &= |(\alpha_1 + \alpha_2)C_1(x) - (\alpha_1 h_A + \alpha_2 h_B + \beta_1 h'_A + \beta_2 h'_B)| \\ &\leq \alpha_1 |C_1(x) - h_A| + \alpha_2 |C_1(x) - h_B| + |\beta_1 h'_A + \beta_2 h'_B| \\ &\leq (\alpha_1 + \alpha_2)\omega(h) + \frac{4}{27}h(h'_A + h'_B) \leq \omega(h) + \frac{8}{27}h \max\{h'_A, h'_B\}. \end{aligned} \tag{11.8}$$

According to the definition of $\omega(h)$, it can be easily deduced that $\lim_{h\to 0} c_1(x) = C_1(x)$; namely, $c_1(x)$ will uniformly converge to $C_1(x)$ as h approaches vanishing point. That is, the model error, which is defined as the difference between the true elevation value and its interpolated value of a point on 1D, approaches zero when the sample density approaches zero.

11.2.2.2 Bicubic Interpolation for TIN

The derivation of cubic interpolation error is much more complex than linear interpolation for TIN or cubic interpolation for 1D. Consider that A, B, and C are three given nodes of a TIN model, the independent variable coordinates $u_A = (x_A, y_A)$, $u_B = (x_B, y_B)$, and $u_C = (x_C, y_C)$ constitute a triangle ΔABC, and the corresponding elevation values of its endpoints are h_A, h_B, and h_C. The cubic interpolation function based on this TIN model is constructed below.

Suppose one point $u_P = (x_P, y_P)^T$ has fallen onto the plain of the TIN ΔABC, as shown in Figure 11.5. How can the approximate elevation of P be obtained by cubic interpolation based on the terrain information (the elevation and derivative) of the triangle's nodes? For cubic interpolation, the details of the procedure are given below:

1. First make three lines parallel to the sides of a triangle of TIN through P, respectively, and the intersection marked as D, E, F, G, H, I in turn.
2. Taking D, E as the endpoints, a cubic polynomial can be constructed by cubic interpolation on the two points, based on their elevations and slopes. The variable coordinates of point P located in the interval of $\overrightarrow{DE}$ can be obtained easily.

   Similarly, the cubic polynomials of the segments $\overrightarrow{FG}$ and $\overrightarrow{HI}$ can be constructed, and the other interpolations of the elevation of P obtained. Neither one nor the other can be accepted or rejected cursorily, and therefore their average is the most persuasive choice.

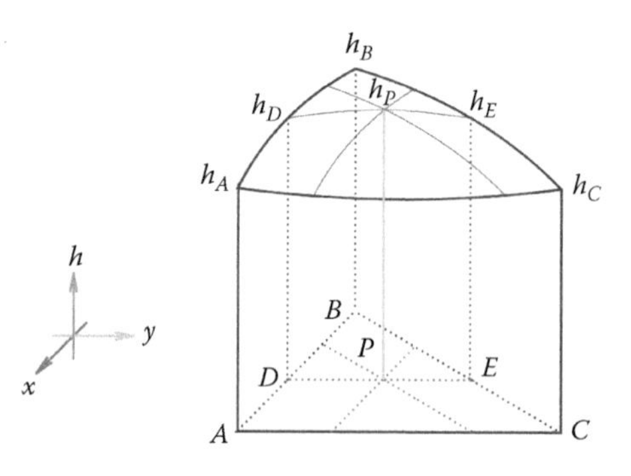

FIGURE 11.5
Cubic interpolation for TIN.

3. In order to obtain the elevation of the boundary points D, E, F, and so forth, the cubic interpolation was also conducted using the three nodes of the TIN triangle.

   Take the cubic interpolation on segment $\overrightarrow{AB}$ and $\overrightarrow{DE}$, for example, to demonstrate the derivation process.

$$c_{AB}(u) = \alpha_1 h_A + \alpha_2 h_B + \beta_1 h'_{\overrightarrow{AB}} + \beta_2 h'_{\overrightarrow{BA}}$$

hereinto

$$\alpha_1 = \left(1 + 2\frac{\|u - u_A\|}{\|u_B - u_A\|}\right)\left(\frac{\|u_B - u\|}{\|u_B - u_A\|}\right)^2$$

$$\alpha_2 = \left(1 + 2\frac{\|u_B - u\|}{\|u_B - u_A\|}\right)\left(\frac{\|u - u_A\|}{\|u_B - u_A\|}\right)^2$$

$$\beta_1 = \left(\|u - u_A\|\right)\left(\frac{\|u_B - u\|}{\|u_B - u_A\|}\right)^2$$

$$\beta_2 = \left(\|u_B - u\|\right)\left(\frac{\|u - u_A\|}{\|u_B - u_A\|}\right)^2$$

$$c_{DE}(x) = \alpha_1 h_D + \alpha_2 h_E + \beta_1 h'_{\overrightarrow{DE}} + \beta_2 h'_{\overrightarrow{ED}}$$

hereinto

$$\alpha_1 = \left(1 + 2\frac{\|u - u_D\|}{\|u_E - u_D\|}\right)\left(\frac{\|u_E - u\|}{\|u_E - u_D\|}\right)^2$$

$$\alpha_2 = \left(1 + 2\frac{\|u_E - u\|}{\|u_E - u_D\|}\right)\left(\frac{\|u - u_D\|}{\|u_E - u_D\|}\right)^2$$

$$\beta_1 = \left(\|u - u_D\|\right)\left(\frac{\|u_E - u\|}{\|u_E - u_D\|}\right)^2$$

$$\beta_2 = \left(\|u_E - u\|\right)\left(\frac{\|u - u_D\|}{\|u_E - u_D\|}\right)^2 \tag{11.9}$$

We can also easily construct the cubic interpolation c_{BC}; it can also easily be constructed by a process similar to that for c_{AB}, in which $h_D = c_{AB}(u_D)$, $h_E = c_{BC}(u_E)$.

4. By linear interpolation, the slope along direction $\overrightarrow{DE}$ can be obtained.

$$h'_{\overrightarrow{DE}} = \frac{\|u_B - u_D\|}{\|u_A - u_B\|} h'_{A\overrightarrow{DE}} + \frac{\|u_A - u_D\|}{\|u_A - u_B\|} h'_{B\overrightarrow{DE}},$$

$$h'_{\overrightarrow{ED}} = \frac{\|u_C - u_E\|}{\|u_B - u_C\|} h'_{B\overrightarrow{ED}} + \frac{\|u_B - u_E\|}{\|u_B - u_C\|} h'_{C\overrightarrow{ED}}.$$

Here $h'_{A\overrightarrow{DE}}$ means the directional derivative along $\overrightarrow{DE}$ at point A and the other notations similarly designed.

5. Therefore,

$$c_T\left(u_P\right)=\frac{1}{3}\left(c_{DE}\left(u_P\right)+c_{FG}\left(u_P\right)+c_{HI}\left(u_P\right)\right).$$

Because cubic interpolation is appropriate only for a plane problem, whereas the DEM discussed here is in three-dimensional space for convenience, $u_A = (x_A,y_A)^T$ and $u - u_A$ have been replaced with their Euclidean form.

Similarly, the cubic interpolations on the other segments, such as $\overrightarrow{BC}$, can also be constructed using the same process as that for c_{AB} and c_{DE}.

Suppose that $C_T(u)$ is the true function of $c_T(u)$ defined on the TIN model, and assume $C_T(u) \in C_{\Delta ABC}$. For any $u_P = (x_P,y_P) \in \Delta ABC$, set $h = \max\{|\overrightarrow{AB}|,|\overrightarrow{BC}|,|\overrightarrow{AC}|\}$; then

$$\begin{aligned}
&\left|C_T\left(u\right)-c_{DE}\left(u_P\right)\right|=\left|\left(\alpha_1+\alpha_2\right)C_T\left(u\right)-\left(\alpha_1 h_D+\alpha_2 h_E+\beta_1 h'_{\overrightarrow{DE}}+\beta_2 h'_{\overrightarrow{ED}}\right)\right|\\
&\le\alpha_1\left|C_T\left(x\right)-h_D\right|+\alpha_2\left|C_T\left(x\right)-h_E\right|+\left\|\beta_1 h'_{\overrightarrow{DE}}+\beta_2 h'_{\overrightarrow{ED}}\right\|\\
&\le\left(\alpha_1+\alpha_2\right)\omega\left(h\right)+\frac{4}{27}h\left(\left\|h'_{\overrightarrow{DE}}\right\|+\left\|h'_{\overrightarrow{ED}}\right\|\right)\le\omega\left(h\right)+\frac{8}{27}h\max\left\{\left\|h'_{\overrightarrow{DE}}\right\|,\left\|h'_{\overrightarrow{ED}}\right\|\right\}\\
&\le\omega\left(h\right)+\frac{8}{27}h\max\left\{\left\|h'_{\overrightarrow{ADE}}\right\|,\left\|h'_{\overrightarrow{BDE}}\right\|,\left\|h'_{\overrightarrow{BED}}\right\|,\left\|h'_{\overrightarrow{CED}}\right\|\right\}\le\omega\left(h\right)+\frac{8}{27}hh'_{\max}.
\end{aligned}$$

Hereinto

$$h'_{\max}=\max_{\substack{\theta\in\left[0,2\pi\right]\\ u\in\Delta ABC}}\left\{\left\|\left(\cos\theta\frac{\partial}{\partial x}+\sin\theta\frac{\partial}{\partial y}\right)C_T\left(u\right)\right\|\right\}.$$

Consequently,

$$\begin{aligned}
\left|C_T\left(u\right)-c_T\left(u_P\right)\right|&=\left|C_T\left(u\right)-\left(\frac{1}{3}\left(c_{DE}\left(u_P\right)+c_{FG}\left(u_P\right)+c_{HI}\left(u_P\right)\right)\right)\right|\\
&\le\frac{1}{3}\left(\left|C_T\left(u\right)-c_{DE}\left(u_P\right)\right|+\left|C_T\left(u\right)-c_{FG}\left(u_P\right)\right|+\left|C_T\left(u\right)-c_{HI}\left(u_P\right)\right|\right)\\
&\le\omega\left(h\right)+\frac{8}{27}hh'_{\max}.
\end{aligned}$$

According to the definition of $\omega(h)$, it can easily be deduced that $\lim_{h\to 0} c_T(x)=C_T(x)$; namely, $c_T(x)$ will uniformly converge to $C_T(x)$ as the sampling density h approaches the vanishing point. That is, the error, which is defined as the difference between the true elevation value and its interpolated value of a point on TIN, approaches zero as the sample density approaches zero.

11.2.2.3 Bicubic Interpolation for Rectangle

Bicubic interpolation for a rectangle can be perceived as a two-step process: (1) y is assumed fixed so that terrain $z(x,y)$ becomes a one-variable function only of x, and the interpolation will be conducted along the x direction only. (2) x is then swapped with y in similar fashion to interpolate the terrain along the orthogonal direction of the anterior.

As shown in Figure 11.6, here A, B, C, and D are considered four given nodes of a rectangle model, where the independent variable coordinates $u_A = (x_1,y_1)$, $u_B = (x_2,y_1)$, $u_C = (x_2,y_2)$, and $u_D = (x_1,y_2)$ constitute a rectangle ($ABCD$)); the corresponding elevation values of its endpoints are h_A, h_B, h_C, and h_D. Referring to Fritsch and Carlson (1980), the cubic interpolation function based on this rectangle model is constructed below, where the following interpolation conditions need to be satisfied:

$$\begin{cases} c_R(x_i,y_i) = h(x_i,y_i) \\ \dfrac{\partial}{\partial x} c_R(x_i,y_i) = \dfrac{\partial}{\partial x} h(x_i,y_i) \\ \dfrac{\partial}{\partial y} c_R(x_i,y_i) = \dfrac{\partial}{\partial y} h(x_i,y_i) \\ \dfrac{\partial^2}{\partial x \partial y} c_R(x_i,y_i) = \dfrac{\partial^2}{\partial x \partial y} h(x_i,y_i) \end{cases}$$

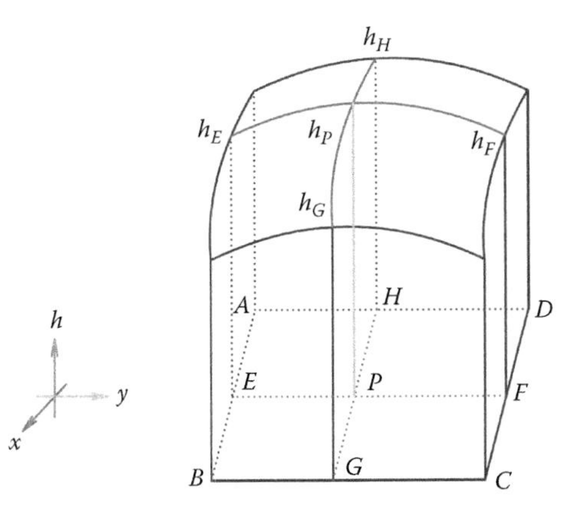

FIGURE 11.6
Cubic interpolation for rectangle.

The first condition of the bicubic interpolation method means that the elevations of the sampling data should be accepted by the interpolation function. The second and third conditions indicate the slopes along the x coordinate axis and the y coordinate axis, respectively, should be utilized. The last one, a mixed partial derivative along both the x axis and the y axis, guarantees the smoothness of the entire terrain. The complete unique expression can be obtained and checked that it satisfies the above conditions.

$$
\begin{aligned}
c_R(x,y) = {} & \sum_{i=1}^{2}\sum_{j=1}^{2} h(x_i,y_j)\alpha_i(x)\alpha_j(y) + \sum_{i=1}^{2}\sum_{j=1}^{2} \frac{\partial}{\partial x} h(x_i,y_j)\beta_i(x)\alpha_j(y) \\
& + \sum_{i=1}^{2}\sum_{j=1}^{2} \frac{\partial}{\partial y} h(x_i,y_j)\alpha_i(x)\beta_j(y) + \sum_{i=1}^{2}\sum_{j=1}^{2} \frac{\partial^2}{\partial x \partial y} h(x_i,y_j)\beta_i(x)\beta_j(y),
\end{aligned}
\tag{11.10}
$$

where

$$
\alpha_1(x) = \left(1 + 2\frac{x - x_1}{x_2 - x_1}\right)\left(\frac{x_2 - x}{x_2 - x_1}\right)^2,
$$

$$
\alpha_2(x) = \left(1 + 2\frac{x_2 - x}{x_2 - x_1}\right)\left(\frac{x - x_1}{x_2 - x_1}\right)^2,
$$

and

$$
\beta_1(x) = (x - x_1)\left(\frac{x_2 - x}{x_2 - x_1}\right)^2,
$$

$$
\beta_2(x) = (x_2 - x)\left(\frac{x - x_1}{x_2 - x_1}\right)^2.
$$

From inequalities (11.6) and (11.7), we know that $\alpha_1 + \alpha_2 = 1$, which implies

$$
\sum_{i=1}^{2}\sum_{j=1}^{2} \alpha_i(x)\alpha_j(y) = 1,
$$

and

$$|\beta_i| \le \frac{4}{27}h.$$

Let

$$M_1 = \max_{u \in (ABCD} \left\{ \left\| \frac{\partial h}{\partial x}(u) \right\|, \left\| \frac{\partial h}{\partial y}(u) \right\| \right\}$$

and

$$M_2 = \max_{u \in (ABCD} \left\{ \left\| \frac{\partial^2 h}{\partial x \partial y}(u) \right\| \right\}.$$

Suppose that $C_R(u)$ is the true function of $c_R(u)$ defined on the rectangle model, and assume $C_R(x) \in C_{(ABCD)}$. For any $u_P = (x_P, y_P)^T \in (ABCD)$, set $h = \max\{|\overline{AB}|, |\overline{BC}|\}$; then

$$\begin{aligned}
|C_R(u) - c_R(u)| \le & \sum_{i=1}^{2}\sum_{j=1}^{2} |C_R(u) - h(x_i, y_j)| \alpha_i(x)\alpha_j(y) \\
& + \left| \sum_{i=1}^{2}\sum_{j=1}^{2} \frac{\partial}{\partial x} h(x_i, y_j)\beta_i(x)\alpha_j(y) \right| \\
& + \left| \sum_{i=1}^{2}\sum_{j=1}^{2} \frac{\partial}{\partial y} h(x_i, y_j)\alpha_i(x)\beta_j(y) \right| \\
& + \left| \sum_{i=1}^{2}\sum_{j=1}^{2} \frac{\partial^2}{\partial x \partial y} h(x_i, y_j)\beta_i(x)\beta_j(y) \right| \\
& \le \left(\sum_{i=1}^{2}\sum_{j=1}^{2} \alpha_i(x)\alpha_j(y) \right) \omega(h) + \left| \sum_{i=1}^{2} \frac{\partial}{\partial x} h(x_i, y_j)\beta_i(x) \right| \\
& + \left| \sum_{j=1}^{2} \frac{\partial}{\partial y} h(x_i, y_j)\beta_j(x) \right| + \left| \sum_{i=1}^{2}\sum_{j=1}^{2} \frac{\partial^2}{\partial x \partial y} h(x_i, y_j)\beta_i(x)\beta_j(y) \right| \\
& \le \omega(h) + \frac{16}{27} M_1 h + \frac{64}{729} M_2 h^2.
\end{aligned}$$

According to the definition of $\omega(h)$, it can easily be deduced that $\lim_{h\to 0} c_R(x) = C_R(x)$; namely, $c_R(x)$ will uniformly converge to $C_R(x)$ as h approaches vanishing point. That is, the model error, which is defined as the difference between the true elevation value and its interpolated value of a point on the rectangle, is approaching zero as the sample density approaches zero.

Thus far, all the convergence analyses for both bilinear and bicubic interpolation over the spatial objects 1D, TIN, and rectangle have been obtained based only on an assumption of practical terrain continuity and the definition of $\omega(h)$. All findings, no matter which interpolation method is adopted, are that the interpolated model error will approach vanishing point as the sample density h approaches zero.

However, the above conclusions are qualitative. In order to obtain a more precise and quantitative model error analysis, a more rigorous mathematical constraint should be guaranteed. To achieve this, the model error, specifically the upper bounds of the model error, are derived below.

11.3 Model Error of Bicubic Interpolation Method

The interpolation model error R_P of a point P over a spatial object, including 1D, TIN, and rectangle, is defined as the difference between the true elevation function and the interpolated elevation function derived from the source data, which are assumed to be error-free. For each patch of a terrain, there exist locations where the maximum difference between the interpolation function and the corresponding true terrain occurs. Let $|R_j|$ denote the maximum difference of the patch j; then the truncated interpolation error (truncated error) for point P over the whole spatial object will be bounded by the maximum $\max\{R_j\}$ among all patches of terrain, that is, $|R_P| \le \max\{R_j\}$. In this study, the focus is on the model error, the truncated error, of the nonlinear interpolation method, specifically the bicubic interpolation method for TIN, and the rectangle model interpolation. For the derivation of model error for linear interpolations, readers may refer to Hu et al. (2009).

11.3.1 Cubic Interpolation for 1D

Introducing the truncated error of cubic interpolation for the 1D method, suppose that $C_1(x)$ $(x_A \le x \le x_B)$ is the true function of the cubic interpolation function $c_1(x)$ $(x_A \le x \le x_B)$. If it is assumed that $C_1(x) \in C^4[x_A, x_B]$, the following can be obtained (Cheney and Kincaid, 2013):

$$\left|R_1(x)\right| = \left|C_1(x) - c_1(x)\right| \le \frac{1}{4!} M_4 \left(x - x_A\right)^2 \left(x - x_B\right)^2 \le \frac{M_4}{384} h^4, \tag{11.11}$$

where

$$M_4 = \max_{x_A \le x \le x_B} \left| C^{(4)}(x) \right|$$

and

$$h = x_B - x_A.$$

By (11.11), it could also be apparently found that the truncated error for 1D is closely associated with the sampling density (h). This relationship can be described as follows.

The truncated error of cubic interpolation for 1D method tends to become smaller with a higher-density sampling. The conclusion here is an alternative but quantitative result confirming the result that was achieved in Section 11.2. However, it is necessary to be aware that the required condition of the true terrain surface in deriving (11.11) $C_1(x) \in C^4[x_A, x_B]$ may, in practice, be hard to achieve.

11.3.2 Cubic Interpolation for TIN

The truncated error of bicubic interpolation for the TIN method is now discussed. Suppose that $C_T(x,y)$ is the true function of the bicubic interpolated $c_T(x,y)$ on TIN model. Assuming that $h = \max\{|\overrightarrow{AB}|, |\overrightarrow{BC}|, |\overrightarrow{AC}|\}$ (see Figure 11.5) and

$$M_n = \max_{p+q=n} \max_{\Delta ABC} \left\| \frac{\partial C_T}{\partial x^p \partial y^q} \right\|,$$

let M'_n denote the maximum norm of the n-order directional derivation (Hu et al., 2009), that is,

$$\begin{aligned} M'_n = \max_{\gamma_1, \gamma_2, \cdots, \gamma_n} \max_{\Delta ABC} & \left\| \left(\cos\gamma_1 \frac{\partial}{\partial x} + \sin\gamma_1 \frac{\partial}{\partial y} \right) \left(\cos\gamma_2 \frac{\partial}{\partial x} + \sin\gamma_2 \frac{\partial}{\partial y} \right) \right. \\ & \left. \cdots \left(\cos\gamma_n \frac{\partial}{\partial x} + \sin\gamma_n \frac{\partial}{\partial y} \right) C_T(x,y) \right\| \\ \le M_n \max_{\gamma_1, \gamma_2, \cdots, \gamma_n} & \left\{ \left(|\cos\gamma_1| + |\sin\gamma_1| \right) \left(|\cos\gamma_2| + |\sin\gamma_2| \right) \cdots \left(|\cos\gamma_n| + |\sin\gamma_n| \right) \right\}. \end{aligned}$$

Special cases of the above formula are

a. $M_1' = M_1 \max_{\gamma}\left(|\cos\gamma| + |\sin\gamma|\right) \le \sqrt{2}M_1$
b. $M_2' \le 2M_2$
c. $M_4' \le 4M_4$

Suppose H_D, H_E, and H_P are the true elevations of the interpolated values h_D, h_E, and h_P, respectively, and denote $\tilde{c}_{DE}(u)$ to be the cubic interpolation function based on the two true elevations of the nodes H_D and H_E.

Then from Equation 11.9,

$$\tilde{c}_{DE}(u) = \alpha_1 H_D + \alpha_2 H_E + \beta_1 H'_{\overline{DE}} + \beta_2 H'_{\overline{ED}} \tag{11.12}$$

is obtained, where

$$\alpha_1 = \left(1 + 2\frac{\|u-u_D\|}{\|u_E-u_D\|}\right)\left(\frac{\|u_E-u\|}{\|u_E-u_D\|}\right)^2,$$

$$\alpha_2 = \left(1 + 2\frac{\|u_E-u\|}{\|u_E-u_D\|}\right)\left(\frac{\|u-u_D\|}{\|u_E-u_D\|}\right)^2,$$

$$\beta_1 = \left(\|u-u_D\|\right)\left(\frac{\|u_E-u\|}{\|u_E-u_D\|}\right)^2,$$

$$\beta_2 = \left(\|u_E-u\|\right)\left(\frac{\|u-u_D\|}{\|u_E-u_D\|}\right)^2.$$

And the bicubic interpolation function based on the two interpolated elevations of the nodes h_D and h_E is

$$c_{DE}(x) = \alpha_1 h_D + \alpha_2 h_E + \beta_1 h'_{\overline{DE}} + \beta_2 h'_{\overline{ED}} \tag{11.13}$$

in which

$$\alpha_1 = \left(1 + 2\frac{\|u-u_D\|}{\|u_E-u_D\|}\right)\left(\frac{\|u_E-u\|}{\|u_E-u_D\|}\right)^2,$$

$$\alpha_2 = \left(1 + 2\frac{\|u_E - u\|}{\|u_E - u_D\|}\right)\left(\frac{\|u - u_D\|}{\|u_E - u_D\|}\right)^2,$$

$$\beta_1 = \left(\|u - u_D\|\right)\left(\frac{\|u_E - u\|}{\|u_E - u_D\|}\right)^2,$$

$$\beta_2 = \left(\|u_E - u\|\right)\left(\frac{\|u - u_D\|}{\|u_E - u_D\|}\right)^2.$$

Replacing

$$h_P = \frac{1}{3}\left(c_{DE}\left(u_P\right) + c_{FG}\left(u_P\right) + c_{HI}\left(u_P\right)\right)$$

with $h_P = c_{DE}(u_P)$ confirms that the truncated error will be no more affected following this exchange.

Therefore, according to (11.12), (11.13), and the absolute value inequality, we have

$$\begin{aligned} \left|\tilde{c}_{DE}\left(u_P\right) - c_{DE}\left(u_P\right)\right| \le \alpha_1\left|H_D - h_D\right| + \alpha_2\left|H_E - h_E\right| \\ +\beta_1\left|H'_{\overline{DE}} - h'_{\overline{DE}}\right| + \beta_2\left|H'_{\overline{ED}} - h'_{\overline{ED}}\right|. \end{aligned} \tag{11.14}$$

Since h_D and h_E are obtained by cubic interpolation on the segments of $\overrightarrow{AB}$ and $\overrightarrow{BC}$, the following inequality is established based on inequality (11.11):

$$\left|H_D - h_D\right| \le \frac{M'_4}{384}h^4 \text{ and } \left|H_E - h_E\right| \le \frac{M'_4}{384}h^4.$$

As described above, the slopes along direction $\overrightarrow{DE}$ are obtained by linear interpolation. From the truncated error of linear interpolation for 1D, which is listed in Table 11.1,

$$\left|H'_{\overline{DE}} - h'_{\overline{DE}}\right| \le \frac{1}{8}M_2h^2 \text{ and } \left|H'_{\overline{ED}} - h'_{\overline{ED}}\right| \le \frac{1}{8}M_2h^2.$$

TABLE 11.1

Comparison of Error Bound of the Bilinear and Bicubic Interpolations

Spatial Objects	Bilinear Interpolation	Bicubic Interpolation
1D	$\lvert R_1\rvert \le \frac{1}{8}M_2h^2$	$\lvert R_1\rvert \le \frac{M_4}{384}h^4$
TIN	$\lvert R_T\rvert \le \frac{3}{8}M_2h^2$	$\lvert R_T\rvert \le \frac{1}{27}M_2h^3 + \frac{1}{48}M_4h^4$
Rectangle	$\lvert R_R\rvert \le \frac{1}{4}M_2h^2 + \frac{1}{64}M_2h^4$	$\lvert R_R\rvert \le \frac{13}{192}M_4h^4$

Combining the above formulas, together with the properties of the parameters used in the cubic interpolation $\alpha_1 + \alpha_2 = 1$ and

$$\lvert\beta_i\rvert \le \frac{4}{27}h.$$

Equation 11.14 becomes

$$\begin{aligned}\lvert\tilde{c}_{DE}(u_P)-c_{DE}(u_P)\rvert &\le(\alpha_1+\alpha_2)\frac{M_4'}{384}h^4+\frac{1}{54}M_2h^3+\frac{1}{54}M_2h^3\\ &\le\frac{1}{27}M_2h^3+\frac{1}{96}M_4h^4.\end{aligned}$$

H_P represents the true elevation and $\tilde{c}_{DE}(u_P)$ represents the bicubic interpolated value based on the true elevations H_D, H_E and true slopes $H'_{\overline{DE}}$, $H'_{\overline{ED}}$. According to (11.11) and (c), also obtained is

$$\lvert H_P-\tilde{c}_{DE}(u_P)\rvert \le \frac{M_4'}{384}h^4 \le \frac{1}{96}M_4h^4.$$

According to the above two inequalities and the absolute value inequality, we have

$$\begin{aligned}\lvert H_P-h_P\rvert &= \lvert(H_P-\tilde{c}_{DE}(u_P))+(\tilde{c}_{DE}(u_P)-c_{DE}(u_P))\rvert\\ &\le\lvert H_P-\tilde{c}_{DE}(u_P)\rvert+\lvert\tilde{c}_{DE}(u_P)-c_{DE}(u_P)\rvert\\ &\le\frac{1}{27}M_2h^3+\frac{1}{48}M_4h^4.\end{aligned}$$

Therefore, the truncated error of bicubic interpolation for a TIN is then given by

$$\left|R_T\left(x,y\right)\right| \le \frac{1}{27}M_2h^3 + \frac{1}{48}M_4h^4. \tag{11.15}$$

11.3.3 Bicubic Interpolation for Rectangle

Consider the truncated error of bicubic interpolation for the rectangle method and suppose that $C_R(x,y)$ is the true interpolation function for rectangle $c_R(x,y)$. We assume that $h = \max\{|\overrightarrow{AB}|, |\overrightarrow{BC}|\}$ and

$$M_n = \max_{p+q=n}\max_{(ABCD}\left\|\frac{\partial C_R}{\partial x^p \partial y^q}\right\|.$$

In estimating the bicubic interpolation error for the rectangle approach, that is, error of the function (11.10), assuming $C_R \in C^4(\Omega)$, the following conclusion can be drawn (Carlson and Fritsch, 1985):

$$\left|C_R\left(x,y\right) - c_R\left(x,y\right)\right| \le \frac{1}{384}\left\{\left\|\frac{\partial^4 f}{\partial x^4}\right\| + 24\left\|\frac{\partial^4 f}{\partial x^2 \partial y^2}\right\| + \left\|\frac{\partial^4 f}{\partial y^4}\right\|\right\}h^4. \tag{11.16}$$

The above inequality can be simplified as

$$\left|R_R\left(x,y\right)\right| \le \frac{13}{192}M_4h^4. \tag{11.17}$$

Based on the model error study of the cubic interpolation for the 1D, TIN, and rectangle methods, together with findings on the model error for linear interpolation for the 1D, TIN, and rectangle methods by Hu et al. (2009), we came up with the summary listed in Table 11.1.

From Table 11.1, interpolation error depends on two factors: M_n and h. Here, M_n is essentially a descriptor of terrain morphology, and h describes sampling density. Furthermore, from Table 11.1, it can be seen that bilinear truncated errors are all of magnitude $O(h^2)$, while truncated errors of the bicubic interpolation are of $O(h^4)$, except for TIN, which is $O(h^3)$. The reason that the error bounder for TIN is large is due to the nature of the TIN model itself. The temporary middle points D, E, and so forth, are used to generate the bicubic curve. Their elevations and slope gradients, which are approximated by the cubic and linear interpolations, respectively, are not given explicitly in advance because they are not from the original source data. However, it can be concluded that the bicubic interpolation results of the 1D, TIN, and rectangle methods are more accurate than those interpolated by the bilinear method, when the original source sampling points are error-free.

Comparing the errors derived from the linear interpolated DEM surfaces with those from bicubic interpolations presented in this chapter, the following conclusion can be reached. Higher-order interpolation methods for DEM will produce more accurate results than those produced by linear interpolation. However, it should be pointed out that, for bicubic interpolation, the terrain slope must be provided in advance since it is impossible to construct a cubic polynomial simply based on two elevations. Although the bicubic interpolation method can provide a higher-accuracy DEM, it is still necessary to understand that it may be difficult or inconvenient to obtain required slope gradient information from the source data directly. In this sense, it is inevitable that other terrain information will be excavated, such as the slope gradient, which may still be calculated by the finite difference method when interpolating a DEM surface, either for one patch or the whole terrain, constituting by many patches.

11.4 Numerical Experiments

In this section, the theoretical findings derived in Sections 11.2 and 11.3 are validated by using the numerical experimental method. The experiments consist of two parts: (1) experience based on mathematical surfaces and (2) experience based on real-world DEM data sets.

The root mean square error (RMSE) value will be used as a measure of model error for bilinear and bicubic interpolation methods in constructing three types of spatial objects by the 1D, TIN, and rectangle methods. The difference between true values and the corresponding interpolated values is expressed in RMSE terms and then averaged over all samples. The square root of the average is then further taken. Since the errors are squared before they are averaged, an RMSE gives a relatively high weight to large errors, and this is thus very useful when large errors are undesirable.

11.4.1 Experiments Based on Mathematical Surfaces

In order to verify the theoretical results obtained in Section 11.3, an experience based on the mathematical surfaces was first designed. With mathematical surfaces, the true values of any points (including checking points) on the surfaces can be obtained easily. The availability of true values for a large number of points is essential for error assessment.

The predefined standard surfaces were designed and generated to estimate model error for the two bilinear and bicubic interpolation methods, for three types of spatial objects: 1D, TIN, and rectangle. The model error is estimated based on the following approaches: (1) graphic visualization and (2) quantitative analysis by an RMSE indicator. A Gaussian synthetic surface function (Zhou and Liu, 2004) was selected for the generation of the mathematical surfaces where the function is defined by

$$z = 3(1-x)^2 e^{-x^2-(y+1)^2} - 10\left(\frac{x}{5} - x^3 - y^5\right)e^{-x^2-y^2} - \frac{1}{3}e^{-(x+1)^2-y^2}. \tag{11.18}$$

For 1D, $z = z(x,0)$, and for TIN and rectangle, $z = z(x,y)$. Terrain information is first assigned, including elevations and slope gradients of m points for randomly generating mathematical surfaces based on (11.18). For 1D, the mathematical surface is a curve, and for TIN and rectangle, the mathematical surfaces are controlled by the assigned terrain information. Next, by applying the bilinear and bicubic interpolation methods, respectively, the interpolated surfaces for 1D, TIN, and rectangle are constructed correspondingly. Finally, n checking points are randomly generated from the interpolated surface for the region under consideration. Their elevations are compared with true values derived from the mathematical surface using (11.18). The RMSE is then computed according to

$$RMSE = \left(\frac{1}{n}\sum_{i=1}^{n}(z_i - h_i)^2\right)^{\frac{1}{2}} \tag{11.19}$$

Figures 11.7 through 11.9 provide a visual comparison of the two interpolation methods, and the numerical RMSE-based results are listed in Table 11.2.

As can be seen from the three figures, the surfaces generated by the bicubic interpolation method are more similar to the original mathematical surface, while the left surface is less similar to the mathematical surface. They are very coarse and discontinuous in comparison with the original smooth and continuous mathematical surface.

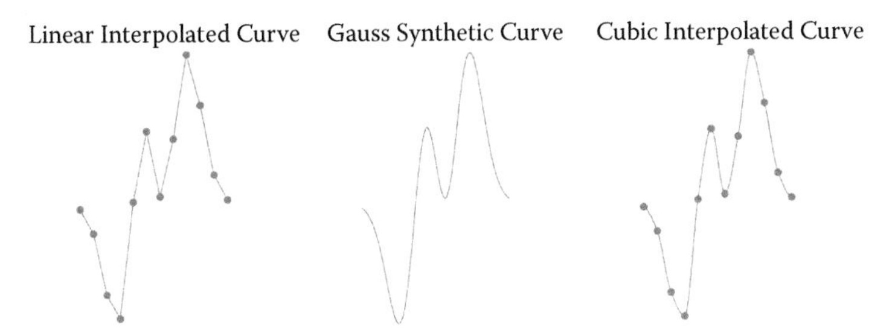

FIGURE 11.7
Test model error by comparing hte interpolated curves for 1D with the Gauss synthetic curve ($m = 12$, $n = 2000$).

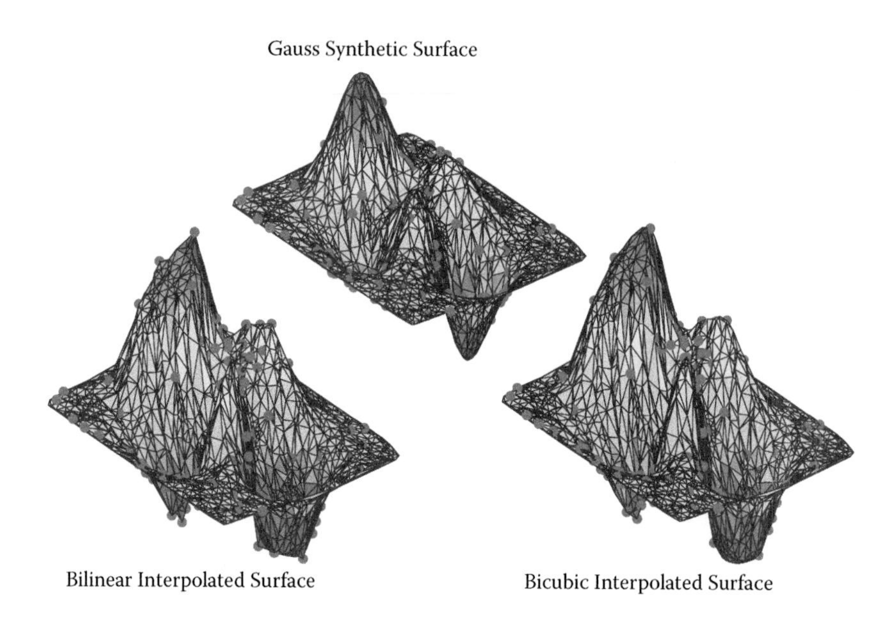

FIGURE 11.8 (See color insert.)
Test model error comparing the interpolated surfaces for TIN with the Gauss synthetic surface ($m = 150$, $n = 1819$).

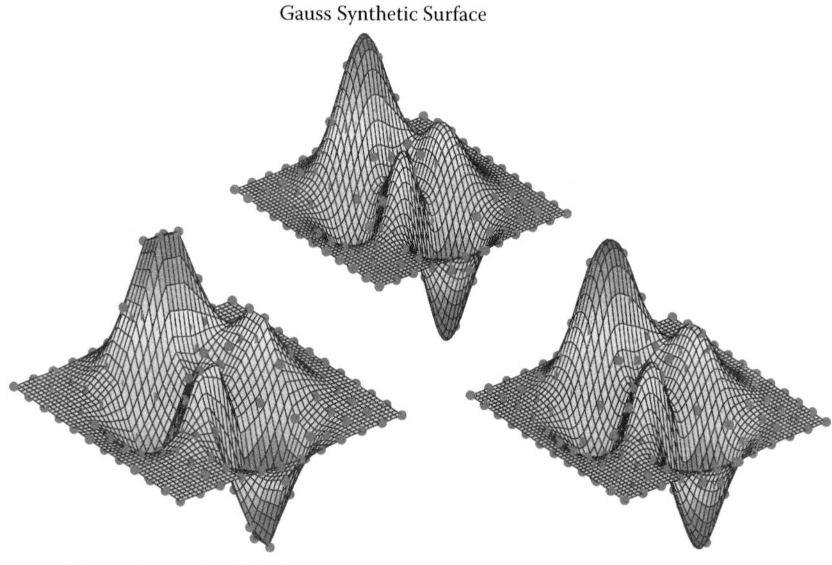

FIGURE 11.9 (See color insert.)
Test model error comparing the interpolated surfaces for rectangle with the Gauss synthetic surface ($m = 144$, $n = 2401$).

TABLE 11.2

Comparison of RMSE between Bilinear and Bicubic Interpolations

Type of Objects	Bilinear Interpolation (m)	Bicubic Interpolation (m)
1D	0.2270	0.1243
TIN	0.3072	0.1958
Rectangle	0.2658	0.0745

As can be seen from Table 11.2, the RMSE errors in the left column, which are generated by the bilinear interpolation method, are significantly greater than those generated by the bicubic interpolation method. These experimental results are consistent with the theoretical findings of Section 11.3.

Compared with the 1D and TIN results, the bicubic interpolated surface error for a rectangle is much smaller (1/4~1/3) than that of the bilinear method. The theoretical results listed in Table 11.1 can explain this, since the model error magnitude is $O(h^4)$ for both 1D and rectangle, while it is $O(h^3)$ for TIN. This means that the upper bound of the model error for a 1D and a rectangle is theoretically much smaller than that of TIN in bicubic interpolation. The model error of the bicubic interpolated curve for 1D is, however, not much smaller than the model error for the bicubic interpolated surface for TIN, as indicated in Table 11.2, because the number of the generated points for 1D was too small ($m = 12$) in this experiment for a precise comparison between the model errors of linear interpolated curves and the model errors of cubic interpolated curves to be derived.

In the above study, the numbers of generated points (m) and checking points (n) were fixed when analyzing the model errors for bilinear and bicubic interpolation models. To verify the authors' theoretical findings in Sections 11.2 and 11.3, the trend and relationship between the model errors of different interpolated surfaces as affected by the sampling interval h are discussed. Here, the interpolated surface error is described by the RMSE value as defined by (11.19), with n randomly selected checking points. This differs from the interpolated surfaces derived from the mathematical surface base. In practice, the maximum sampling interval h_{max} used in the model error formulas may not be easy to obtain. Alternatively, the number of samples (m generated points) is used as an indicator of h_{max}.

As the sampling data were randomly selected from the interpolated surface, the profiles plotted in the above three figures cannot smoothly present the relationship between the DEM interpolation accuracy (RMSE) and the sampling density.

From Figures 11.10 through 11.12, the following two conclusions can be drawn. First, RMSE of the interpolated curve for the 1D, TIN, and rectangle cases practically vanishes with a very large sampling number m. This means that the RMSE model error approaches zero when the maximum sampling interval (h_{max}) approaches zero. This experimental conclusion verifies

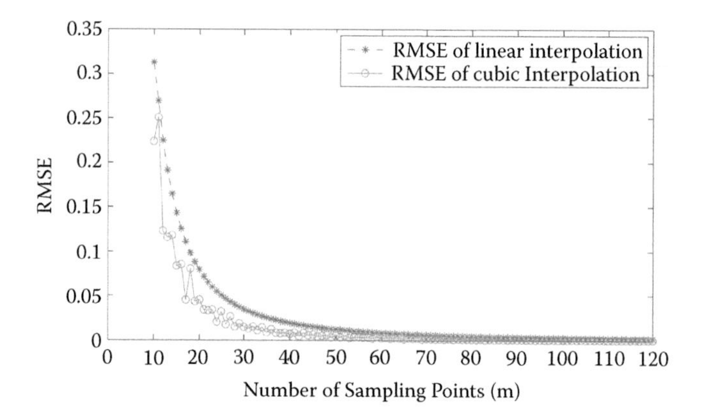

FIGURE 11.10
A comparison of model error of the interpolated curve from the linear and cubic interpolations for 1D.

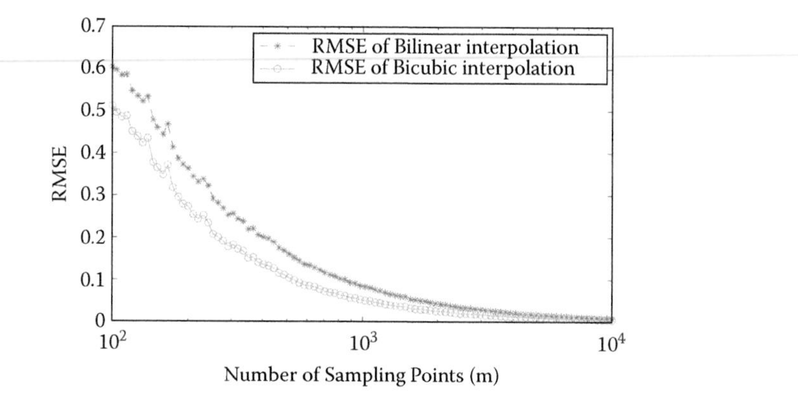

FIGURE 11.11
A comparison of model error of the interpolated surface from the bilinear and bicubic interpolations for TIN.

the same theoretical findings of Section 11.2. Second, the RMSE curves for bicubic interpolation are always beneath the bilinear interpolation curves (Figures 11.10 through 11.12). This means that the model error for a bicubic interpolated surface is smaller than that for a bilinear interpolated surface for the 1D, TIN, and rectangle cases. This experimental result again verifies the theoretical findings in Section 11.3. This experimental result, with the varying sampling number approaching infinity, is even more convincing than the early experimental result based on a fixed sampling number.

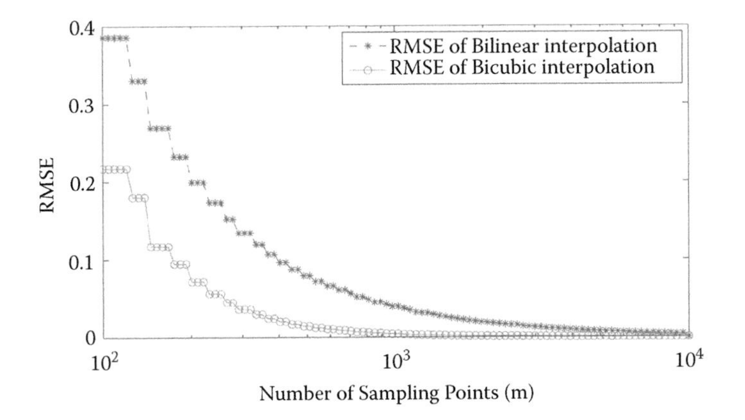

FIGURE 11.12
A comparison of model error of the interpolated surface from the bilinear and bicubic interpolations for rectangle.

11.4.2 Experiments Based on Practical Data Sets

To demonstrate the applicability of the theoretical findings to the practical real-world data, the following experiment was designed based on practical DEM data used in industry. Shanxi province (left part of Figure 11.13) was chosen for our experimental study, and three test areas were further identified representing the three typical terrain types: plain, hill, and mountain.

Sample region A is located within 33.56°–33.837°N and 107.476°–107.754°E. It lies in the Qianlong Mountains, in the southern part of Shanxi province, which is mountainous. Sample region B is located within 34.348°–34.626°N and 108.13°–108.408°E and is situated on the central Shanxi plain. Sample region C is located within 36.064°–36.342°N and 109.492°–109.77°E. It is a loess hill and gully area in northern Shanxi, a hilly topography. All three sample regions are illustrated in Figure 11.13, and their characteristics are described in Table 11.3.

An Advanced Spaceborne Thermal Emission and Reflection Radiometer (ASTER) Global Digital Elevation Model (GDEM), produced by the Ministry of Economy, Trade and Industry of Japan and the U.S. National Aeronautics and Space Administration (NASA), was used for the study. The ASTER GDEM is in GeoTIFF format with geographic latitude and longitude coordinates and a 1 arc-sec (30 m) grid of elevation postings.

It is referenced to the WGS84/EGM96 geoid. Preproduction estimated accuracies for this global product were 20 m at 95% confidence for vertical data and 30 m at 95% confidence for horizontal data. When calculating the terrain gradient value, the projection was converted to UTM/WGS84. Three topographic surfaces were extracted measuring 1000 × 1000 pixels from the original data set to act as the test data.

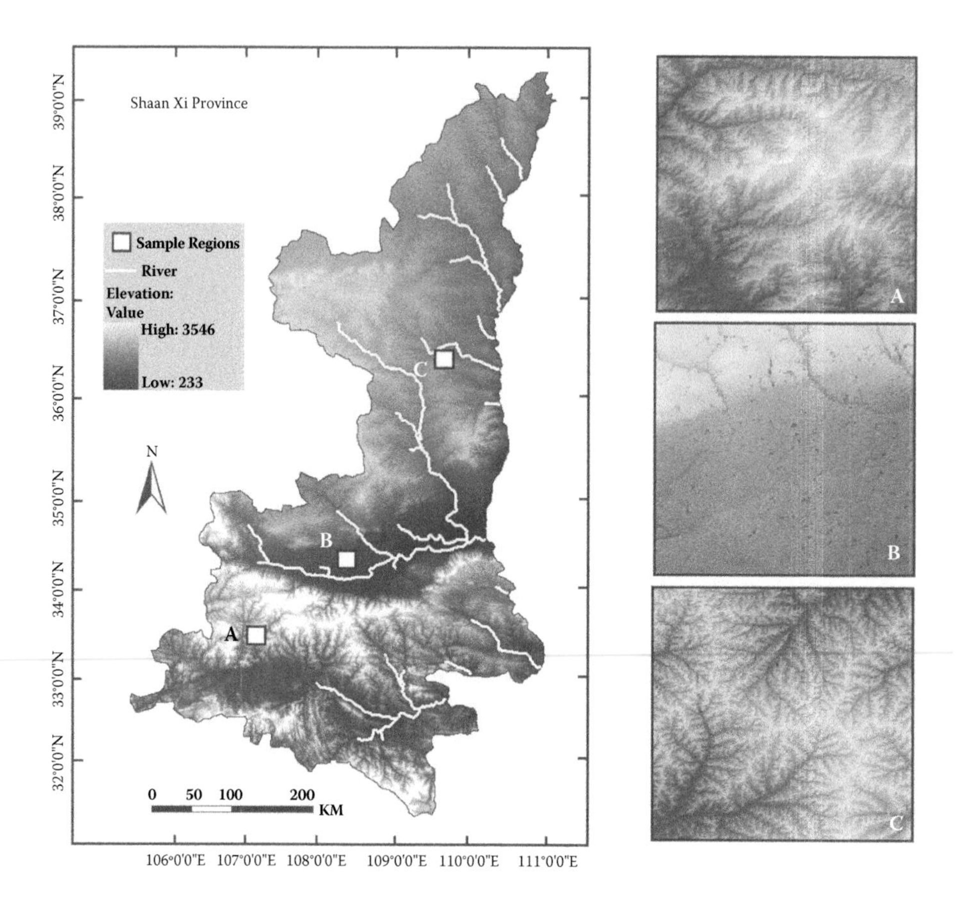

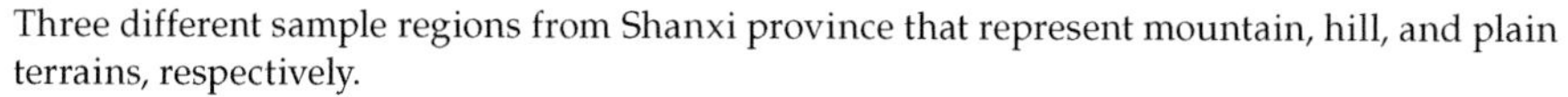

FIGURE 11.13
Three different sample regions from Shanxi province that represent mountain, hill, and plain terrains, respectively.

As for the experiment based on the mathematical surface, m points were first chosen randomly from the sample data set. The interpolated surfaces were then generated using the bilinear and bicubic interpolation methods, respectively. Finally, n points were randomly generated in the region discussed, and their elevations compared with the corresponding true value from the samples. The RMSE was used as a measurement of the model error and is summarized in Table 11.4.

From the results listed above, it can be seen that, utilizing RMSE for measuring model error, the bicubic interpolated DEM is smaller than that for the bilinear interpolated DEM. This result is consistently valid for the plain region, hill region, and mountain region, no matter which elementary object, TIN or rectangle, is adopted. These experimental findings fully verify the theoretical findings of Section 11.3. Besides, another conclusion can also be

TABLE 11.3

Characteristics of the Topographic Surfaces of the Studied Area

Terrain Descriptive Statistics	High Mountain	Hill	Plain
Min (m)	986	1005	242
Max (m)	3052	1464	401
Average elevation (m)	1907.95	1265.57	338.41
Hmax – Hmin (m)	2066	459	159
Standard deviation of elevation (m)	404.17	78.28	11.26
Elevation coefficient of variation (%)	21.18	6.19	3.33
Min slope (°)	0	0	0
Max slope (°)	78	68.28	52.41
Average slope (°)	24.05	15.01	3.27
Standard deviation of slope (°)	10.30	8.32	2.49
Slope coefficient of variation (%)	42.81	55.44	76.11

Note: All surfaces are grid DEMs composed of 10^6 points and with 30 × 30 m spacing.

TABLE 11.4

Comparison of RMSE of Bilinear and Bicubic Interpolations (m)

Terrain Types	Spatial Objects	m	n	Bilinear	Bicubic
Plain	TIN	221,382	411,121	1.2698	1.0607
	Rectangle	62,500	931,509	1.5773	1.2899
Hill	TIN	221,114	410,872	3.7673	2.8575
	Rectangle	62,500	931,509	4.7284	3.0795
Mountain	TIN	221,098	411,125	5.3769	4.2097
	Rectangle	62,500	931,509	6.4467	4.3236

acquired that the interpolated errors (indicated by the RMSEs) become larger with the increasing of terrain complexity. Meanwhile, the bicubic interpolation method demonstrates great superiority compared to the bilinear one, as the interpolated DEM by the formal model possesses much higher accuracy than its counterpart by the latter approach.

11.5 Conclusions

Model error, that is, the truncated error of the bicubic interpolated DEM surface, has been investigated in this research. First, the relationship between the DEM error and the sampling density for both bilinear interpolation and bicubic interpolation methods for the elementary spatial objects—1D, TIN, and rectangle—has been qualitatively explored by convergence analysis. The conclusion indicates that the interpolated error, represented by the upper bound of the truncated error, is approaching the vanishing point, along with a decrease of the sampling interval. Such a convincing finding is achieved via rigorous analytical derivations, which strongly support our intuitive feel about the relationship between the interpolated DEM error and the sampling density.

Second, the truncated error bound of the bicubic interpolation method has been explored quantitatively for the first time. This is an advance compared with earlier studies of the model error of the bilinear interpolation method. The model error for a nonlinear (bicubic) interpolated DEM surface investigated in this study, together with the research outcome for a linear interpolated DEM surface from the earlier study (Hu et al., 2009), forms a full picture of model error for interpolated DEM surfaces. These results can serve as a reference for interpolation model selection in DEM production practice.

Both the theoretically derived results and the experimental studies have demonstrated that model error (truncated error) of the bicubic interpolated DEM surface is smaller than the error for a bilinear interpolated DEM surface, making the assumption that the sample points are error-free; namely, the interest centrality has been inclined upon the interpolated DEM error.

Although the propagated error can be very small during the data capturing process, especially by modern technology, such as LIDAR, a further study is recommended for the analysis of overall error, including both model error and the error propagated from the original sample data. Another direction for follow-up extension of this study is the derivation of DEM errors by using other nonlinear interpolation models taking the spatial correlation into consideration, such as the Kriging interpolation approach.

Acknowledgments

The authors would like to thank Teng Zhong from Hong Kong University for providing the ASTER GDEM data and Mrs. Elaine Anson of Hong Kong Polytechnic University for her careful proofreading and language polishing. Mr. Bin Wang is the beneficiary of a doctoral grant from the AXA Research Fund. The authors also acknowledge funding support from the Ministry of Science and Technology of China (project no. 2012AA12A305, 2012BAJ15B04), and the Hong Kong Polytechnic University (1-ZVBA).

References

Aguilar, F.J., Aguilar, M.A., Agüera, F. 2007. Accuracy assessment of digital elevation models using a non-parametric approach. *International Journal of Geographical Information Science* 21, 667–686.

Aguilar, F.J., Aguilar, M.A., Agüera, F., Sánchez, J. 2006. The accuracy of grid digital elevation models linearly constructed from scattered sample data. *International Journal of Geographical Information Science* 20, 169–192.

Arun, P. 2013. A terrain-based hybrid approach towards DEM interpolation. *Annals of GIS* 19, 245–252.

Carlisle, B.H. 2005. Modelling the spatial distribution of DEM error. *Transactions in GIS* 9, 521–540.

Carlson, R. and Fritsch, F. 1985. Monotone Piecewise Bicubic Interpolation. *SIAM J. Numer. Anal.*, 22(2), 386–400.

Cheney, E.W., Kincaid, D. 2013. *Numerical Mathematics and Computing.* 7th ed. Brooks/Cole/Cengage Learning, Boston.

Fritsch, F.N., Carlson, R.E. 1980. Monotone piecewise cubic interpolation. *SIAM Journal on Numerical Analysis* 17, 238–246.

Gao, J. 1997. Resolution and accuracy of terrain representation by grid DEMs at a micro-scale. *International Journal of Geographical Information Science* 11, 199–212.

Guo, Q., Li, W., Yu, H., Alvarez, O. 2010. Effects of topographic variability and LIDAR sampling density on several DEM interpolation methods. *Photogrammetric Engineering and Remote Sensing* 76, 701–712.

Hu, P., Liu, X.H., Hu, H. 2009. Accuracy assessment of digital elevation models based on approximation theory. *Photogrammetric Engineering and Remote Sensing* 75, 49–56.

Huang, Y. 2000. Evaluation of information loss in digital elevation models with digital photogrammetric systems. *Photogrammetric Record* 16, 781–791.

Kyriakidis, P.C., Goodchild, M.F. 2006. On the prediction error variance of three common spatial interpolation schemes. *International Journal of Geographical Information Science* 20, 823–855.

Li, Z., Zhu, C., Gold, C. 2010. *Digital Terrain Modeling: Principles and Methodology.* CRC Press, Boca Raton, FL.

Liu, X., Sherba, J. 2012. Accuracy assessment of LiDAR-derived DEM based on approximation theory. Presented at Proceedings of the Annual Conference of American Society of Photogrammetry and Remote Sensing, ASPRS, Bethesda, MD, 2012.

Oksanen, J., Sarjakoski, T. 2006. Uncovering the statistical and spatial characteristics of fine toposcale DEM error. *International Journal of Geographical Information Science* 20, 345–369.

Shi, W., Wang, B., Tian, Y. 2014. Accuracy analysis of digital elevation model relating to spatial resolution and terrain slope by bilinear interpolation. *Mathematical Geosciences*, 1–37.

Shi, W.Z., Li, Q.Q., Zhu, C.Q. 2005. Estimating the propagation error of DEM from higher-order interpolation algorithms. *International Journal of Remote Sensing* 26, 3069–3084.

Zhou, Q.M., Liu, X.J. 2004. Analysis of errors of derived slope and aspect related to DEM data properties. *Computers and Geosciences* 30, 369–378.

Zhu, C., Shi, W., Li, Q., Wang, G., Cheung, T., Dai, E., Shea, G. 2005. Estimation of average DEM accuracy under linear interpolation considering random error at the nodes of TIN model. *International Journal of Remote Sensing* 26, 5509–5523.

12

Automatic Method of Inspection for Deformation in Digital Aerial Imagery Based on Statistical Characteristics

Yaohua Yi, Yuan Yuan, Hai Su, and Mingjing Miao

School of Printing and Packaging, Wuhan University, Wuhan, People's Republic of China

CONTENTS

ABSTRACT The geometric distortion in push-broom digital aerial imagery can be rectified using data from an inertial measurement unit (IMU). However, inaccuracies in the IMU data cause undulant, wavelike twist deformations in the push-broom digital aerial images after geometric calibration, which directly influences the authenticity and liability of the images and their practical application. At present, the diagnosis of image deformation depends mainly on the subjective judgment of a human being, which costs a great deal of time and manpower. In this chapter, an automatic method of inspection for deformation in digital aerial imagery based on statistical characteristics is proposed to inspect for the distortions caused by inaccurate IMU data. Because the undulant, wavelike deformations in an image regularly feature the displacement of pixels in the same direction, many wave curves in the image appear in the same direction after geometric calibration. Therefore, in this method, the positions of the wave curves in the image are located by the extreme points of curvature of the contour lines, and the wavelike deformations can then be judged automatically through the distribution statistics of the open directions of the wave curves. The specific method implemented can be described as follows. First, the edges of the image are detected with a Canny edge detector, and the vector contour lines are obtained by tracing the edges and fitting them with the cubic spline curve method. The extreme points of curvature of the contour lines are then calculated, and some of these points are determined to be the vertices of the wave curves by judging the positional relationships between each extreme point and the points around them, thus constituting a vertex set. The perpendicular directions of the tangents of the vertices are then used as the directions of the wave curves, and the direction histograms of all of the wave curves in the image are obtained by statistical analysis. Finally, the existence of the deformation phenomenon in the image due to the inaccurate IMU data can be determined on the basis of whether the directions of the wave curves are centralized in a certain direction. The experimental results show that the automatic method of inspection for deformation presented in this chapter can effectively detect with 95% accuracy the deformation in digital aerial images caused by inaccurate IMU data.

12.1 Introduction

With advances in technology and widely practical applications, the technology for POS (global positioning system [GPS]/inertial measurement unit [IMU])–based aerial photogrammetry has become more mature and is gradually being applied in each field of remote sensing. It is often adopted to correct the geometric distortions in push-broom aerial remote sensing images using POS data to improve image quality. The POS system includes GPS and IMU data. A linear interpolation calculation is always carried out for IMU data with poor accuracy to satisfy the requirement of geometric calibration. However, interpolation errors result in regional pixel displacement in the same direction of the image after geometric calibration (Figure 12.1). This distortion phenomenon severely affects the geometric accuracy of the digital aerial images and their normal use.

In practical applications, examination of the deformation in aerial images without POS data is often required. For the deformations caused by

FIGURE 12.1
Schematic of the testing image with distortions caused by inaccurate data from an inertial measurement unit. Original image. Testing image after geometric calibration, with regional pixel displacement in the same direction.

inaccurate IMU data, this examination currently depends mostly on the subjective judgment of a human being, which costs a great deal of time and manpower. In engineering practice, image definition is the most common index by which to inspect the deformation of aerial images. However, image definition is more useful as a quality indicator by which to evaluate the sharpness of edges in images, rather than as a method of detecting image deformations. Furthermore, although the definition will decrease if deformations exist in the images, the reduction will not be significant if the deformation is particularly strong. In other words, the common distortions are not sufficiently sensitive to the definition of the aerial imagery. At the same time, many other factors contribute to some extent to the changes in image definition. Therefore, as an inspection index, the definition has less utility in the inspection of deformations caused by limited precision. Effective methods are still lacking in the relevant literature both at home and abroad in terms of a method of inspection for deformation in digital aerial images.

12.2 Method of Inspection for Deformation in Digital Aerial Imagery Based on Statistical Characteristics

This chapter proposes an effective method of solving this problem. Some concepts need to be explained in detail for the benefit of description. The curve segments of the texture contours that approximate a parabolic section in the image are defined as squiggles, and the opening directions of the squiggles act as their directions. In addition, the deformation phenomenon of the imagery described in this chapter refers only to that caused by inaccurate IMU data.

A great deal of experimental analysis led to a key breakthrough regarding this issue. As illustrated above, insufficiently accurate IMU data cause pixel displacement in a regular direction in a portion of the aerial image, and noticeable wavelike deformations arise at the edges of the images in the corresponding directions. Although normal images also contain some texture contours that approximate squiggles, the aerial photographs that record information on the ground and its texture contours are usually irregular. As a result, the directions of the squiggles are uniformly distributed in all directions instead of being oriented in a certain direction, as is usually the case in the distorted images. Therefore, it is possible to inspect the deformation phenomenon by examining the statistical directions of the squiggles and constructing a histogram, which is the main idea of the method proposed in this chapter.

The core problem in constructing the histogram of the directions concerns the method of identifying the squiggles from the complex digital aerial images. Common methods of searching for and matching a specified graphic or template generally include the Hough transform, template matching, and invariant moment matching. However, because of the various shapes of the squiggles of push-broom digital aerial images that lack sufficient IMU accuracy, it is difficult to describe functions and establish templates for them; hence, the conventional matching methods are not applicable or suitable.

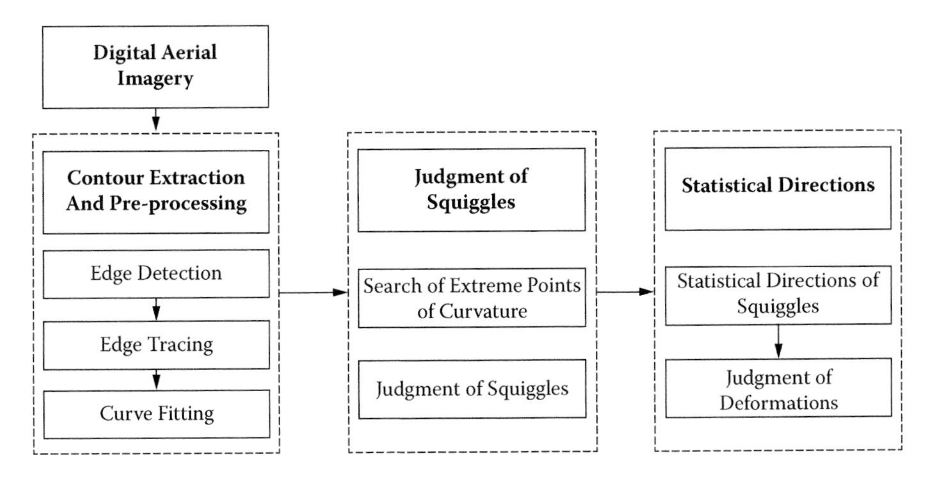

FIGURE 12.2
Workflow of method of inspection for deformation in digital aerial imagery based on statistical characteristics.

Through subjective observation and objective test analysis, most squiggles have been found to be similar to parabolas, and their corresponding directions are always consistent with the normal directions of the vertices of the squiggles. Moreover, the approximate vertices are essentially located at the minimum curvature points. Therefore, the positions of the squiggles can be determined by searching the extreme points of curvature of the contour lines in the image. With a solution for the problem of squiggles, a method of inspection for deformation in digital aerial imagery based on statistical characteristics can be obtained in the following descriptions.

The basic workflow of the method is shown in Figure 12.2 and mainly involves three procedures. The first procedure is contour extraction and preprocessing, which consists of three steps: edge detection, edge tracing, and curve fitting. In the second procedure, judgment of the squiggles is conducted by searching the contour lines for extreme points of curvature and then determining whether they are the vertices of the squiggles. Finally, in the third procedure, the normal directions of the vertices of the squiggles are calculated, and the distributions of those directions are obtained through statistical analysis. The deformation of the aerial imagery can then be determined.

12.2.1 Contour Extraction and Preprocessing of Digital Aerial Imagery

To locate the squiggles, it is necessary to extract all of the contours from an image and judge each of them one by one. The contour extraction of an image is composed of three steps: edge detection, contour tracing, and curve fitting.

12.2.1.1 Edge Detection

The current common methods of edge detection consist of the Roberts operator, the Sobel operator, the Prewitt operator, the Laplacian operator, and the Canny operator. Of these, the Canny operator is based on the idea of optimization and has the advantages of resistance to noise and accurate positioning. It was therefore adopted for edge detection in this chapter because of its high degree of accuracy. The steps of the Canny edge detection are described as follows, and the extracted contours of the test image are shown in Figure 12.3:

1. Smooth the image with a Gaussian filter of specific standard deviations σ, thus reducing factors such as the noise caused by light intensity and the temperature and quality of the sensor.
2. For each pixel in the image, compute the amplitude and direction of the gradients through the finite difference of first-order partial derivatives.
3. Conduct nonmaximum suppression processing to identify all edges. Partition the gradients according to the directions, and search the local maximum pixels. Then set the value of the nonmaximum pixel to zero to sharpen the edges of the image.
4. Judge the pixels using thresholds T_1 and T_2, where $T_1 < T_2$. A pixel with a value greater than T_2 is called a strong-edge pixel, and a pixel with a value between T_1 and T_2 is called a weak-edge pixel.

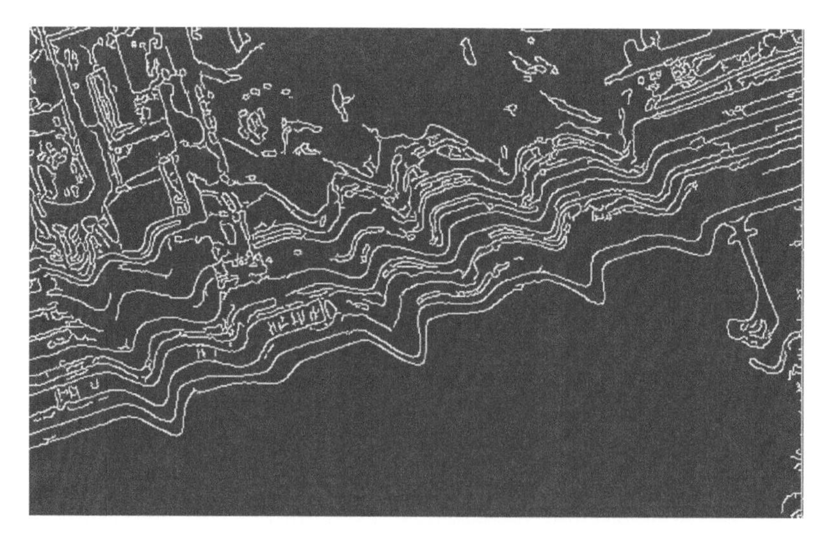

FIGURE 12.3
Contours extracted with the Canny method of edge detection.

5. Set the weak-edge pixels whose eight neighboring pixels have a strong edge as strong-edge pixels, and proceed with binary processing based on whether the edge pixels are strong or weak.

12.2.1.2 Contour Tracing

After the edges of the image have been acquired, the contour tracing process is performed and the data of the contour lines are saved in the form of a list to facilitate the location of squiggles. The concrete steps of contour tracing are described as follows, and the contour lines after contour tracing can be seen in Figure 12.4:

1. Perform a sequential search of all pixels in the image from left to right and bottom to top; locate the boundary pixel at the bottom left corner and designate it as the start point.
2. From the first boundary point, the initial search direction is defined as left top. If the pixel at the left top is black, it is the boundary point. Otherwise, the direction will be rotated 45° clockwise until the first black pixel appears.
3. Set the first black pixel as the new boundary point, and rotate the current search direction 90° counterclockwise. Apply the same method to search for the next black pixel until returning to the original boundary point or until no pixels are available.
4. Save the traced contours into the list and label them as tracked.
5. Repeat the first step until the image has nonlabeled contours.

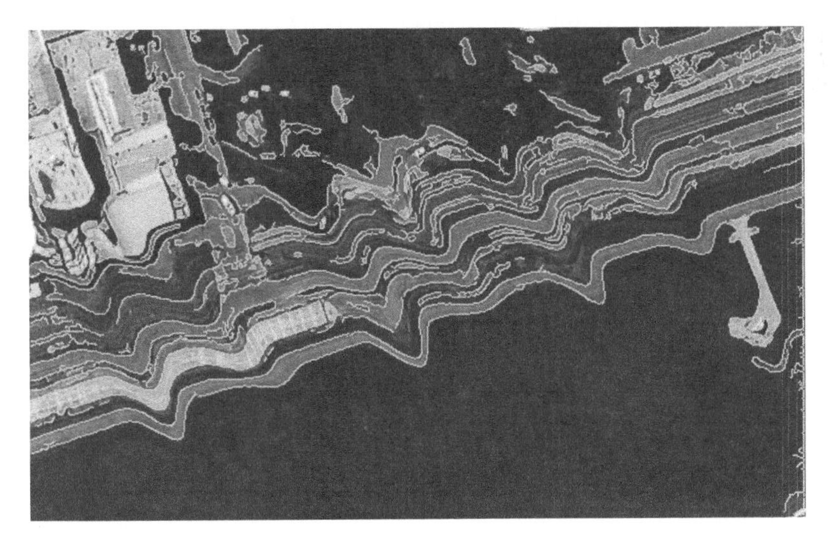

FIGURE 12.4 (See color insert.)
Contour lines after contour tracing.

FIGURE 12.5 (See color insert.)
Curve fitting results with cubic spline function.

12.2.1.3 Curve Fitting

The extracted contours are discrete and cannot be applied directly into the graphic computing; the use of smooth and derivable spline functions to fit the contours of the image can solve this problem. The curve fitting process of a spline function is also able to eliminate the noise for the contours of the image and facilitate judgment of the squiggles of the image at the same time.

Suppose that the number of pixels in the current contour is n. The functional relationship between the parameter t ($t = 0, 1, 2, \ldots, n - 1$) and the plane curve formed by the contours of the current image is obtained with the cubic spline function fitting method, which is

$$c(t) = (x(t), y(t)) \tag{12.1}$$

where $x(t)$ is the functional relationship between the x-axis of the plane curve and parameter t, and $y(t)$ is the functional relationship between the y-axis of the plane curve and parameter t.

Figure 12.5 shows the curve fitting results with a cubic spline function.

12.2.2 Search for and Judgment of Squiggles in Aerial Imagery

12.2.2.1 Search for Extreme Points of Curvature

In this method, the possible positions of squiggles are determined by searching the extreme point of curvature of each contour in a push-broom digital aerial image. It can then be determined whether those points constitute a squiggle according to the spatial relationship between each extreme point of curvature and the surrounding points.

For the plane curve $c(t) = (x(t), y(t))$ given with parametric expressions, the calculation formula of the curvature K is shown in (12.2):

$$K = \frac{|x'(t)y''(t) - x''(t)y'(t)|}{\left(x'^2(t) + y'^2(t)\right)^{3/2}} \tag{12.2}$$

where $x'(t)$ is the first-order derivative of the functional relationship between the x-axis of the plane curve and parameter t, $y'(t)$ is the first-order derivative of the functional relationship between the y-axis of the plane curve and parameter t, $x''(t)$ is the second-order derivative of the functional relationship between the x-axis of the plane curve and parameter t, and $y''(t)$ is the second-order derivative of the functional relationship between the y-axis of the plane curve and parameter t.

The derivation calculation is applied to both sides of the formula of curvature to obtain the relationship between the first derivative K' of the curvature of the image contours and parameter t.

$$K' = \frac{\left(x'(t)y'''(t) - x'''(t)y'(t)\right)\left(x'^2(t) + y'^2(t)\right)}{\left(x'^2(t) + y'^2(t)\right)^{5/2}} - \frac{-3\left(x'(t)y''(t) - x''(t)y'(t)\right)\left(x'(t)x''(t) + y'(t)y''(t)\right)}{\left(x'^2(t) + y'^2(t)\right)^{5/2}} \tag{12.3}$$

where $x'''(t)$ is the third-order derivative of the functional relationship between the x-axis of the plane curve and parameter t, and $y'''(t)$ is the third-order derivative of the functional relationship between the y-axis of the plane curve and parameter t.

If K' is 0, the pixel corresponding to parameter t_0 is the extreme point of curvature in the contour of the image.

Because the power time of parameter t in the equation $K'(t) = 0$ is greater than 2, the root formula cannot be used directly. Instead, only approximate values are determined with a numerical method. The function $K'(t)$ is smooth and derivable, and both positive and negative values exist around the extreme points of curvature corresponding to the squiggles according to their characteristics. Therefore, the method of bisection can be used to solve the equation. General methods of bisection can only find the simple root and odd multiple roots for the equation, but a single image contour might have multiple different extreme points of curvature. For this reason, the bisection method requires adjustment and modification. The concrete steps after adjustment are as follows.

Suppose that the number of the pixels in the current processing contour is n, $a_0 = 0$, $b_0 = n$. For $k = 0, 1, \ldots, m$, perform the following steps:

1. Compute

$$x_k = \frac{a_k + b_k}{2}.$$

2. If $b_k - a_k \le \varepsilon$ or $|f(x_k)| \le \eta$, then set $s \approx x_k$ and record the approximate value s in the root list. Take the k from the list to be processed and replace the current k. x_k and b_k from the list are defined as the processing regions (a_k, b_k); otherwise, go to step 3.
3. If $f(a_k)f(x_k) < 0$ and $f(x_k)f(b_k) > 0$, make $a_{k+1} = a_k$ and $b_{k+1} = x_k$; if $f(a_k)f(x_k) > 0$ and $f(x_k)f(b_k) < 0$, make $a_{k+1} = x_k$ and $b_{k+1} = b_k$; and if $f(a_k)f(x_k) < 0$ and $f(x_k)f(b_k) < 0$, make $a_{k+1} = a_k$ and $b_{k+1} = x_k$. Then save the current k, x_k, and b_k in the list to be processed.
4. If $k = m$, output the unsuccessful iteration information m times. Take the k from the list to be processed to replace the current k. x_k and b_k from the list are defined as the processing regions (a_k, b_k); otherwise, continue.

12.2.2.2 Judgment of Squiggles

The parameter t corresponding to each extreme point of curvature in the contours of the image is calculated via the adjusted bisection method. However, the extreme points of curvature in the contours are not necessarily the vertices of the squiggles, so it is essential to judge these extreme points one by one.

The waves formed from the squiggles of the digital aerial images possess a certain amplitude and wavelength that is comparatively obvious in the image. The judgment of whether the extreme point and its surrounding points develop noticeable waves should be made on the basis of the squiggles' characteristics. The method used here is explained as follows.

Suppose that the number of the pixels in the contours in which the extreme points of curvature are located is n and that the parameter corresponding to the current judging extreme point is t_0 and the preset judging range of the surrounding points is $(t_0 - 1, t_0 + 1)$, where 1 is the threshold for the surrounding points.

1. If $t_0 - 1 < 0$ or $t_0 + 1 > n - 1$, which indicates that the extreme points are at the two ends, the number of surrounding points is not sufficient to compose a squiggle and the judgment result is false; otherwise, proceed to the next step.
2. According to the function expression of fitting a curve to the image contours, compute the x-axis and y-axis of P_0, $P_0 - 1$, and $P_0 + 1$ in the contours corresponding to t_0, $t_0 - 1$, and $t_0 + 1$.

FIGURE 12.6 (See color insert.)
The inspected squiggles in an aerial image.

3. Compute the Euclidean distance between $P_0 - 1$ and $P_0 + 1$. If the value is less than threshold T_1, the judgment result is false; otherwise, proceed to the next step.
4. Compute the distance from P_0 to the segment between $P_0 - 1$ and $P_0 + 1$. If the value is less than threshold T_2, the judgment result is false; otherwise, record P_0 as the vertex of the squiggle.
5. Determine all of the extreme points of curvature in the contours of the image according to the above steps and record all of the vertices of the squiggles (Figure 12.6).

12.2.3 Statistical Directions and Distortion Judgment for the Squiggles of Aerial Imagery

12.2.3.1 Statistical Directions of Squiggles

The waves formed from the squiggles and the surrounding points in digital aerial images are not standard parabolas. Consequently, the symmetrical directions of the waves cannot act as the opening directions of the waves in the same way as parabolas. However, the normal directions of the vertices of the squiggles can represent well the directions of the squiggles; therefore, the normal direction of each vertex is determined as the direction of each squiggle.

According to the parametric equation of the squiggles of the image, the rad of the normal directions of the vertices of the squiggles are computed:

$$rad = \text{atan}\left(-\frac{y'(t)}{x'(t)}\right) \tag{12.4}$$

The data rad in the radian system are converted into θ in the degrees system, whose range is [–180, 180]. When $x'(t) = 0$, $y'(t) < 0$, $\theta = -180$; when $x'(t) = 0$, $y'(t) > 0$, $\theta = 180$; then θ is saved into the list.

Each direction of each squiggle is computed one by one according to the above steps.

12.2.3.2 Judgment of Deformations

After the directions of all of the squiggles have been calculated, statistical analysis is carried out to inspect the deformation phenomenon of the image. To determine the statistical directions of the squiggles, the principle first needs to be developed.

Based on the common directions of the squiggles caused by deformation, the angle range of 0°–360° is divided equally into eight intervals of 45°. In the same distorted image, the directions caused by deformation usually appear with paired angles in the form of complementary angles. Therefore, when performing statistical analysis, those angles should be classified as the same characteristic. To facilitate calculation, the threshold range of the arc tangent is adopted as the statistical range, which is [–180, 180]. Division of the statistical range into four direction intervals yields [–22.5, 22.5], [22.5, 67.5], [–67.5, –22.5], and [–90, –67.5] ∪ [67.5, 90], as shown in the shadowed region of Figure 12.7.

Because the tangent values of the complementary angles are the same, the results of the arc tangent are directly used as the data for statistical analysis regardless of the unit conversion of the angle data calculated by formula (12.3). The number of squiggles in each direction interval is counted, and the judgment value R_{max} is then computed by dividing the maximum number

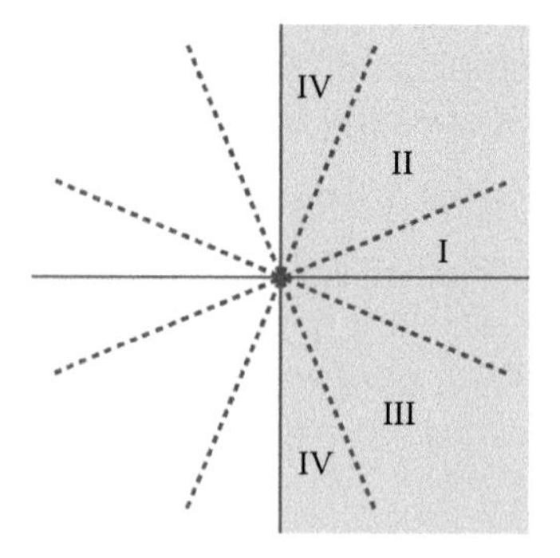

FIGURE 12.7
Schematic of the intervals of statistical directions.

with the most squiggles by the total number of squiggles. Finally, the presence of deformation is determined by comparison of R_{max} with the threshold T_{jud}. The preset T_{jud}, with data in the range (0.25, 1), directly influences the results of judging. The lower the threshold T_{jud}, the easier it is to judge the deformation.

$$Rmax = Num\_max/Num\_total \tag{12.5}$$

where *Num_max* is the number of the squiggles with the most squiggles in the direction interval, and *Num_total* is the total number of squiggles.

12.3 Experiments and Results

Forty push-broom digital aerial images (11,000 × 8,000) were selected to undergo analysis with the method presented here; 10 had deformation. The method of inspection for deformation in digital aerial imagery based on statistical characteristics was applied to these images with the thresholds $T_1 = 6$, $T_2 = 6$, and $T_{jud} = 0.35$.

The results are shown in Table 12.1. Of the 30 images without deformation, 28 were correctly classified as having no distortions and the remaining 2 images were classified as having distortions; deformations were correctly identified in all 10 of the images that had deformation. The average time of calculation for each image was 45 s, and the accuracy rate of the inspection method was 95%. Two of the images had erroneous results because the geomorphic features in the images had distinct squiggles in the same directions.

The test image mentioned above is used as an example; the statistical results are listed in Table 12.2. The highest number of squiggles is 45, and the total number is 110. Therefore, the judgment value R_{max} is 0.41, which is greater than the T_{jud} threshold of 0.35. The result is that the aerial image contains deformation, which is in agreement with the facts.

TABLE 12.1

Results of Inspection of 40 Images

	Statistics		
Image	**Total Number**	**Number Detected**	**Number Not Detected**
Image without deformation			
Image with deformation			0

TABLE 12.2

Results of Inspection of a Test Image

	Statistics	
Direction Interval	Number of Squiggles	R_{max}
[–22.5, 22.5]	39	0.41
[22.5, 67.5]	45	
[–67.5, –22.5]	10	
[67.5, 90] ∪ [67.5, 90]	16	

12.4 Summary

An automatic method of inspection for deformation in digital aerial imagery based on statistical characteristics is proposed for inspection of the deformation caused by inaccurate IMU data to replace manual subjective judgment or unsatisfactory indexes. The method achieved a high degree of accuracy and can be performed with a relatively high speed, which is beneficial in practical applications. However, the texture contours of the images directly affect the statistical results of the corresponding directions of the squiggles. Therefore, the discrimination between the squiggles used for detection of deformation and the contours that belong to the image content will be a focus of future studies. In addition, the differences in the precise extent of the IMU data lead to some changes in the amplitudes and lengths of the squiggles in the images. As a result, the self-adaptation of judgment thresholds on the basis of actual images requires further study before automatic inspection can be realized.

References

1. Lijun Z., Qiu L., Yibin H., Dahai G., and Yueguan Y. Evaluation of GPS/IMU supported aerial photogrammetry. In *IEEE International Geoscience and Remote Sensing Symposium*, IGARSS, 2006, pp. 1512–1514.
2. Xueyou L., Rongjun Z., Yingcheng L., and Heping L. The application of IMU/DGPS-based photogrammetry in large scale map production. *Science of Surveying and Mapping*, 31(1), 60–61, 2006.
3. Shunping J. and Ma D. GPS/IMU based SIFT-matching for aerial remote sensing images. In *Proceedings of 2011 International Conference on Remote Sensing, Environment and Transportation Engineering*, RSETE, Nanjing, 2011, pp. 1257–1260.
4. Lailiang S., Chunxi Z., and Daihong C. An elastic deformation measurement method for helicopter based on double-IMUs/DGPS TRAMS. *Journal of the International Measurement Confederation*, 46(5), 1704–1714, 2013.

5. Ballard D. H. Generalizing the Hough transform to detect arbitrary shapes. *Pattern Recognition*, 13(2), 111–122, 1981.
6. Rafael C. G. and Richard E. W. *Digital Image Processing*. Beijing, 2004 (in Chinese).
7. Xinxing H., Qian L., Li T., and Duoxiang C. Implementation of the push-broom digital aerial photography GPS/IMU data quality automated inspection. *Surveying and Mapping*, 06, 271–273, 2013.
8. Choi M., Daehee W., Sangkyoung S., et al. The effects of using heading measurement during alignment of a low-cost IMU/GPS system. In *26th International Technical Meeting of the Satellite Division of the Institute of Navigation*, ION GNSS, Nashville, TN, 2013, vol. 3, pp. 2175–2179.
9. Ming-Kuei H. Visual pattern recognition by moment invariants. *IRE Transactions on Information Theory*, 8(2), 179–187, 1962.
10. Wei Q., Keyou W., Chuanyong Z., and Jing S. Research on the technique of IMU/GPS assisted ADS80 digital aerial photogrammetry system. *Geospatial Information*, 11(3), 106–108, 2013.
11. Kouibia A., Pasadas M., and Rodriguez M. L. Optimization of parameters for curve interpolation by cubic splines. *Journal of Computational and Applied Mathematics*, 235(14), 4187–4198, 2011.

13

Comparison of Point Matching Techniques for Road Network Matching

A. Hackeloeer,[1] K. Klasing,[1] J. M. Krisp,[2] and L. Meng[3]
[1]*BMW Forschung und Technik GmbH, Munich, Germany*
[2]*Department of Geography, Universität Augsburg, Augsburg, Germany*
[3]*Department of Cartography, Technische Universität München, Munich, Germany*

CONTENTS

ABSTRACT Map conflation investigates the unique identification of geographical entities across different maps depicting the same geographic region. It involves a matching process that aims to find commonalities between geographic features. A specific subdomain of conflation called road network matching establishes correspondences between road networks of different maps on multiple layers of abstraction, ranging from elementary point locations to high-level structures such as road segments or even subgraphs derived from the induced graph of a road network. The process of identifying points located on different maps by means of geometrical, topological, and semantic information is called point matching. This chapter provides an overview of various techniques for point matching, which is a fundamental requirement for subsequent matching steps focusing on complex high-level entities in geospatial networks. Common point matching approaches as well as certain combinations of these are described, classified, and evaluated. Furthermore, a novel similarity metric called the Exact Angular Index is introduced, which considers both topological and geometrical aspects. The results offer a basis for further research on a bottom-up matching process for complex map features, which must rely on findings derived from suitable point matching algorithms. In the context of road network matching, reliable point matches provide an immediate starting point for finding matches between line segments describing the geometry and topology of road networks, which may in turn be used for performing a structural high-level matching on the network level.

13.1 Introduction

Conflation can be seen as the process of identifying geographical entities across different maps depicting the same geographic region, which are then combined to create a new map. According to a definition proposed by Longley et al. (2005), conflation is "the process of combining geographic information from overlapping sources so as to retain accurate data, minimize redundancy, and reconcile data conflicts." A classification approach introduced by Yuan and Tao (1999) divides conflation into *horizontal* (combining neighboring areas) and *vertical* (combining different maps of the same area) conflation. Throughout this chapter, we will focus on vertical conflation, while most point matching techniques are applicable to both types.

In general, three different types of information can be used in the conflation process: *geometrical, topological,* and *semantic.* Geometrical information describes geometric properties of an object, such as the shape of a road segment. Topological information is exposed by the graph structure induced by networks of certain geographical objects, such as roads or rivers. Semantic information can be seen as any kind of information that does not belong to the other two categories (e.g., street names). Both raster image and vector data may be used for conflation. However, different conflation strategies are required depending on the type and direction (raster-to-raster [Christmas et al., 1994], raster-to-vector [O'Donohue, 2010], vector-to-raster [Kovalerchuk et al., 2008], or vector-to-vector [Gösseln, 2005]). Throughout this chapter, we will focus on vector-to-vector pairings of maps.

A specific subdomain of conflation called *road network matching* (Walter and Fritsch, 1999) investigates correspondences between road networks of different maps, which may be established on different levels, ranging from elementary point locations to complex aggregated structures such as sequences of road segments. All mentioned types of information can be considered for each of these levels. A common approach in the domain of road network matching involves a bottom-up matching strategy (Xiong, 2000) that starts with *point matching*, that is, finding relations between point locations. These matching results are then further processed in order to provide a basis for higher-level matches between aggregated structures such as road segments.

This work is concerned with *introducing*, *classifying*, and *evaluating* point matching techniques for road network matching. The overview (Section 13.2) describes the point matching problem in general. Section 13.3 gives a classification of point matching techniques based on the type of information considered and describes the different approaches in detail, including a novel approach named Exact Angular Index. In Section 13.4, the described point matching techniques are evaluated with respect to properties such as accuracy and complexity in a real-world scenario involving maps from different sources such as OpenStreetMap. Section 13.5 discusses the results, which are summarized in Section 13.6. We therefore intend to provide the reader with a quick understanding of the advantages and disadvantages of the presented point matching techniques, which offer a starting point for identifying higher-level matches required within the road network matching process.

13.2 Overview of the Point Matching Problem

Figure 13.1 shows a road map of the village of Moosach, near Munich, Germany, provided by the Bavarian State Office for Survey and Geoinformation, Munich, which is called *ATKIS Basis-DLM*. This area is overlaid with a road map built from geographic data provided by the volunteered geographic information (VGI) project *OpenStreetMap* (OSM).

Topologically, each map consists of a graph (the road network) given by *edges* and *nodes* (vertices), where a node represents a geographical point referenced via its coordinates in a suitable reference system (e.g., WGS-84), and an edge is given by a relation that describes a connection between two nodes. For the purposes of point matching, the graph may be seen as being

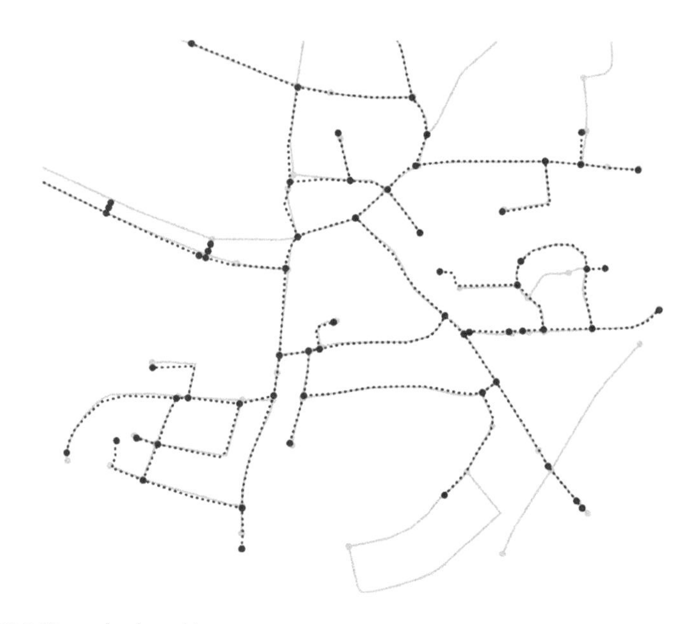

FIGURE 13.1 (See color insert.)
Overlay of two maps of the same region: gray solid, ATKIS; black dashed, OpenStreetMap.

undirected to simplify the process. It should be noted that many map providers insert bivalent nodes (nodes that are only incident to two edges) at locations where attributes change, which are recorded per edge. Occasionally, the terms *0-cells* for nodes, *1-cells* for edges, and subsequently, *2-cells* for polygons consisting of a sequence of edges are used (Rosen and Saalfeld, 1985).

Geometrically, the shape of a road segment represented by an edge in the graph is described via *shape points* (not explicitly shown in the figures), where a sequence of shape points constitutes the geometrical layout of the road segment corresponding to an edge. Like nodes, shape points are defined by their coordinates. Continuous shape geometry is created by employing linear interpolation between shape points.

While it is possible to employ point matching strategies to any data consisting of spatial point coordinates, such as *points of interest* (POIs) or shape points, matching topological nodes in a road network is of special concern for the domain of road network matching. The following box gives a formal definition of solutions to the point matching problem for two road networks with respect to topological nodes.

As can be seen in Figure 13.1, several problems surface when dealing with the point matching problem in real-world scenarios:

1. Topological differences: The two maps do not share the same topology. Rather, roads are present in one map that are missing in the other map. Also, due to a varying level of detail, even roads present in both maps may be modeled with a different number of nodes. In

DEFINITION 13.1: POINT MATCHING FOR TWO ROAD NETWORKS

Given are two undirected simple graphs G_1, G_2 representing a road network: $G_i = (V_i, E_i)$, $i \in \{1,2\}$, where V_i is the set of vertices of graph G_i, and E_i is the edge relation of graph G_i, and $E_i \subseteq (V_i \times V_i)$. Then, a *point matching* is a pair (p_1, p_2), where $p_1 \in V_1$ and $p_2 \in V_2$.

DEFINITION 13.2: SOLUTION TO THE POINT MATCHING PROBLEM FOR TWO ROAD NETWORKS

A *solution* to the point matching problem for two road networks is a relation $S \subseteq (V_1 \times V_2)$, that is, a set of point matchings.

DEFINITION 13.3: LEFT-UNIQUE SOLUTION TO THE POINT MATCHING PROBLEM FOR TWO ROAD NETWORKS

A *left-unique solution* to the point matching problem for two road networks is an injective relation $S_l \subseteq (V_1 \times V_2)$.

DEFINITION 13.4: RIGHT-UNIQUE SOLUTION TO THE POINT MATCHING PROBLEM FOR TWO ROAD NETWORKS

A *right-unique solution* to the point matching problem for two road networks is a functional relation $S_r \subseteq (V_1 \times V_2)$.

DEFINITION 13.5: UNIQUE AND COMPLETE SOLUTIONS TO THE POINT MATCHING PROBLEM FOR TWO ROAD NETWORKS

A *unique solution* to the point matching problem for two road networks is a right-unique and left-unique relation $S_u \subseteq (V_1 \times V_2)$. Note that $S_r \cap S_l$ is a unique solution. If S_u is also surjective and left-total, it is a *complete solution.*

NOTES:

- A solution according to Definition 13.2 represents an M:N type mapping between nodes of V_1 and V_2.
- A left-unique solution according to Definition 13.3 represents a 1:N type mapping between nodes of V_1 and V_2.
- A right-unique solution according to Definition 13.4 represents an N:1 type mapping between nodes of V_1 and V_2.
- A unique solution according to Definition 13.5 represents a 1:1 type mapping between nodes of V_1 and V_2.
- A complete solution only exists if $|V_1| = |V_2|$.

addition, structures such as complex intersections may be modeled differently, resulting in a different placement of nodes.

2. Geometrical differences: Due to varying accuracy in the recorded coordinates, the geographical location of nodes representing the same topological entity may differ between the maps to a great extent. On the other hand, nodes in close proximity do not necessarily imply a topological relationship.
3. Semantic differences: Nodes may carry semantic information, such as the names of incident roads. While semantic similarity of two nodes in different maps often indicates that the same node is referenced, semantic dissimilarity rarely implies the opposite, since semantic attributes, as well as the extent to which they are recorded, vary greatly across different map providers and sources. For example, street names may be spelled differently, and there may also be multiple names for the same street.

In order to deal with these problems, several algorithms have evolved that determine and evaluate *point matching candidates* (i.e., a subset of all point matchings) with respect to certain metrics. In general, a metric is defined as can be seen in Definition 13.6.

Thus, in the context of the point matching problem, a metric is a distance function that assigns a real number to a point matching, where the assigned value expresses the degree of dissimilarity of the two points involved. The distance may be normalized, for example by projecting it onto the interval (0; 1], where 1 corresponds to the lowest possible distance (0, meaning equality) and 0 corresponds to an infinitely high distance. We call such a projection a *score* because it is positively correlated with the expected quality of the matching from the perspective of the according metric. The overall score that an algorithm attributes to a point matching may be a weighted combination of multiple scores from different metrics.

In order to limit the computational effort, point matching algorithms usually only evaluate a small subset of all possible point matches. Most point matches are discarded beforehand due to spatial constraints, since it is

DEFINITION 13.6: METRIC DEFINITION FOR POINT MATCHING

A *metric* on a set $X := V_1 \cup V_2$ is a function $d{:}X \times X \rightarrow \mathbb{R}$, where $\forall x, y, z \in X$; the following conditions must be satisfied:

a. $d(x,y) = 0 \Leftrightarrow x = y$
b. $d(x,y) = d(y,x)$
c. $d(x,z) \leq d(x,y) + d(y,z)$

assumed that the probability of two points representing the same spatial entity quickly becomes extremely low as the distance between the points increases beyond several kilometers. The fact that point matching algorithms, unlike pure graph matching approaches, may rely not only on topological but also on geometrical information thus greatly simplifies the point matching problem, since it reduces the number of candidates that need to be evaluated.

While a complete solution as defined in Definition 13.5 may seem desirable, it only exists for very simple scenarios where the two maps being compared are virtually identical (e.g., very minor map updates). In real-world applications, usually both maps contain several nodes that cannot reasonably be matched to the other map.

Point matching algorithms deliver a set of point matchings, where each point matching is assumed to identify the same geographical entity across both maps. This is done by evaluating the degree of similarity between point matchings close enough to become candidates according to certain metrics, and then selecting those point matchings for the solution that are considered similar enough that they may reasonably represent the same spatial feature (e.g., by applying a global or local threshold on the score). In general, this solution is ambiguous, implying that any point on any of the two maps may be matched to more than one point in the other map. It is possible to establish unique solutions by discarding all matchings related to a node except the highest-rated point matching. However, in special cases, a geographical entity represented by one topological node in one map may be (partially) represented by multiple nodes in the other map (e.g., bivalent nodes, differing level of detail, or complex intersections), so these cases must be recognized and dealt with separately.

13.3 Classification and Description of Point Matching Techniques

Point matching techniques may be based on geometrical, topological, or semantic information, or a combination of these three. Since most map providers do not include semantic attributes for topological nodes and semantic matching techniques referring to incident edges are more appropriate for edge matching, we will not discuss semantic matching in greater detail.

13.3.1 Geometrical Point Matching Techniques

Geometrical point matching techniques only consider geometrical information (i.e., coordinates) for evaluating a point matching. Even though the distance between point coordinates may be calculated in any p-norm, the only metric of practical relevance is the Euclidean distance metric.

13.3.1.1 Pure Euclidean

Obviously, spatial proximity is a strong constraint for the selection of matching candidates. In the Euclidean plane, the distance dist(p,q) of two points $p = (p_x, p_y)$ and $q = (q_x, q_y)$ is given by

$$\text{dist}(p,q) = \sqrt{(p_x - q_x)^2 + (p_y - q_y)^2}$$

While for small geographical regions, Euclidean geometry is a good approximation, a more exact measure for the distance between two points on the surface of the Earth is the great-circle distance, which describes the shortest distance between any two points on the surface of a sphere following a path on the surface. The great-circle distance can be computed using the spherical law of cosines or the numerically better-conditioned Haversine formula (Sinnott, 1984). The approaches presented in this chapter use a Euclidean distance metric applied to a *Universal Transversal Mercator* (UTM) projection of WGS-84 coordinates for calculating distances, as it provides sufficient accuracy and can be computed efficiently.

A naïve approach to conduct Pure Euclidean point matching calculates the Euclidean distance dist(p,q) for each point matching pair $(p_i, q_j) \in (V_1 \times V_2)$, which possesses all properties of a metric function as defined in Definition 13.6. To extract a right-unique solution S_r, each $p_i \in V_1$ is assigned to a $q_j \in V_2$ where $dist(p_i, q_j)$ becomes minimal so that $(p_i, q_j) \in S_r$ and $(p_i, q_{k \neq j}) \notin S_r \forall q_k \in V_2$. In order to gain a left-unique solution S_l, each $q_j \in V_2$ may only be related to a $p_i \in V_1$ where $dist(p_i, q_j)$ becomes minimal so that$(p_i, q_j) \in S_l$ and $(p_{k \neq i}, q_j) \notin S_l \forall p_k \in V_1$. Finally, a unique solution $S_u = S_r \cap S_l$ can be derived.

This approach is obviously inefficient as it evaluates every possible point matching pair. However, by employing a spatial index (e.g., a kd-tree) and only evaluating neighbors within a sufficiently large radius, pure Euclidean matching may be performed efficiently without losing substantial accuracy.

The Euclidean distance may be projected to a score of the interval (0; 1] by using the following formula:

$$\text{score}_{\text{PE}}(p_i, q_j) = \frac{1}{1 + \left(\frac{\text{dist}(p_i, q_j)}{c \cdot r}\right)^2}$$

where $c \in \mathbb{R}^+$ is a correction factor and $r \in \mathbb{R}^+$ is the node search radius. Note that $\lim_{\text{dist} \to \infty} \text{score}(p_i, q_j) = 0$ for any c, r with $c \cdot r \neq 0$.

13.3.2 Topological Point Matching Techniques

Topological point matching techniques employ topological information such as the *valence* (number of incident edges) per node.

13.3.2.1 Node Valence

The valence, or *degree*, of a node in a graph is defined as the number of edges incident to the node, where loops are counted twice.

The *node valence* point matching approach is concerned with the differences in valence found between the two nodes of a point matching. The larger this difference grows, the lower the probability becomes that both nodes reference the same geographical entity. However, minor differences in valence are no guarantee that the nodes are a bad match, since the two maps may differ in actuality or level of detail so that, for example, small roads may only be present in one map. Also, a valence difference of zero does not imply node equality, so that node valence on its own does not qualify as a metric, since it violates condition (a) in Definition 13.6.

The valence difference may be combined with the Euclidean distance to obtain an order in case of equal valence difference. Since point matchings of large geometrical distances are unlikely to represent a proper solution, the search for matching candidates regarding a node may be limited to its neighborhood within a certain radius. Only within this neighborhood does node valence need to be evaluated. The lower the difference of valence is, the higher the assumed similarity of two nodes. If two matching candidate pairs are assigned to the same equivalence class regarding their valence difference, the pair with the lower Euclidean distance is assumed to be more similar. Depending on properties of the maps to be matched, such as node density or dispersion, node valence may require a fine-tuning of the search radius to provide acceptable solutions.

A straightforward approach for calculating a score based on valence difference is reflected by the following formula:

$$\text{score}_{\text{NV}}\left(p_i, q_j\right) = \frac{1}{1 + \left|\text{val}\left(p_i\right) - \text{val}\left(q_j\right)\right|}$$

13.3.3 Combined Geometrical and Topological Point Matching Techniques

Combined geometrical and topological point matching techniques are employed by algorithms that follow both geometrical and topological approaches and combine them in order to achieve better matching results.

13.3.3.1 Spider Index

Rosen and Saalfeld (1985) describe a point matching technique called the *Spider Index*, which overlays a node with a circular eight-sector discretization of 45° angle intervals similar to a compass rose. Each sector corresponds to a single bit within an 8-bit number. A bit is set to true if and only if there is an incident edge that falls into the corresponding sector. Thus, each node p can be described by an 8-bit number with bits $p_{b_1} \cdots p_{b_8}$. For two nodes $p \in V_1$ and $q \in V_2$, the score of the Spider Index is then calculated as

$$\text{score}_{\text{SI}} = \frac{1}{8}\sum_{i=1}^{8} \vartheta\left(p_{b_i}, q_{b_i}\right)$$

where $\vartheta(p_{b_i}, q_{b_i}) = p_{b_i} \leftrightarrow q_{b_i}$ is the binary equivalence function, that is, $\vartheta = 1$ if bits are equal and $\vartheta = 0$ if they are different. Note that we have normalized the score to the interval [0; 1].

Due to the information loss resulting from the quantization, two nodes may still be considered equal if the angles of their incident edges are different within the limits of a sector. Moreover, it may happen that two nodes are not considered equal if the angles of their incident edges are rotated by a tiny degree, but beyond the limits of a sector. Yet, compared to node valence, the Spider Index offers a more accurate measure for node equality, as it not only accounts for topological valence difference, but also geometrical angle difference.

As with node valence, the Spider Index alone does not qualify as a metric, since two matching candidate pairs whose bit difference regarding their Spider Index is zero may still be different. However, in the same way as with node valence, the Spider Index may be turned into a metric by combining it with the Euclidean distance, so that Euclidean distance determines an order where the Spider Index does not discriminate.

13.3.3.2 Exact Angular Index

Here, we introduce a novel similarity metric called the Exact Angular Index (EAI). Like the Spider Index, the EAI aims to find point matching solutions that consider both topological valence and geometrical angle difference of incident edges. However, the EAI does not employ quantization. Rather, the best mapping between the edges of the two nodes of a matching candidate, that is, the mapping that minimizes the angle differences between the vectors derived from the geometrical shapes of the edges, is determined by evaluating all possible edge mappings. Then, a score is calculated based on the sum of minimum angle differences according to the mapping relative to the largest possible sum of angle differences, where differences in valence are counted as the worst possible angle differences.

Formally, the algorithm follows these steps to iteratively assign a score to point matchings $\{(p_i, q_j) | p_i \in V_1, q_j \in V_2\}$ for each p_i with incident edges $E_{p_i} \subseteq E_1$:

- For each incident edge of p_i, calculate the geographical heading, that is, the angle between the vector given by the first linear segment of the edge and true north in the clockwise direction. The result is a heading function $h_{p_i}: E_{p_i} \rightarrow \mathbb{R}$.

- Search for nodes $q_1 \ldots q_n \in V_2$ in G_2 within a fixed radius around the position of p_i. If no surrounding nodes can be found, no matching partner can be assigned to p_i, so the algorithm continues with the next node p_{i+1}.
- For each found node $q_j \in V_2$ with incident edges $E_{q_j} \subseteq E_2$:
  1. Calculate the heading function $h_{q_j}:E_{q_j} \rightarrow \mathbb{R}$.
  2. Calculate the best-mapping function $b_{q_j}^{p_i}:E_{p_i} \rightarrow E_{q_j}$, which determines the optimum mapping from each edge incident to p_i to an edge incident to q_j regarding their angle difference, using h_{p_i} and h_{q_j}. If $|E_{p_i}| > |E_{q_j}|$, there are edges $\overline{E_{pl}} \subseteq E_{p_i}$ that could not be mapped to edges of E_{q_j} and $b_{q_j}^{p_i}(e) = \varnothing \forall e \in \overline{E_{pl}}$.
  3. Calculate the sum of all angle differences s_{all} by adding up the differences between the headings of all best mappings gained from $b_{q_j}^{p_i}$:

$$s_{\text{all}} = \sum_{k=1}^{n} \Delta\left(h_{p_i}(e_k), h_{q_j}\left(b_{q_j}^{p_i}(e_k)\right)\right)$$

     where $1 \leq k \leq n$, $\{e_1, \ldots, e_n\} \subseteq E_{p_i} \setminus \overline{E_{p_l}}$, and $\Delta(\alpha,\beta)$ is the angle difference computed by $\Delta(\alpha,\beta) = \text{mod}(|\alpha - \beta|, 360)$. If Δ is larger than 180, it is subtracted from 360, so that reflex angles are avoided.
  4. If there is a difference in valence between p_i and q_j, add an angle difference of 180° per missing or redundant edge to get the normalized sum of all angle differences s_{norm}:

$$s_{\text{norm}} = s_{\text{all}} + 180 \cdot |\text{val}(p_i) - \text{val}(q_j)|$$

     where val(p) is the valence of node p.
  5. Calculate the largest possible sum of angle differences s_{largest}:

$$s_{\text{largest}} = 180 \cdot \max\left(\text{val}(p_i), \text{val}(q_j)\right)$$

     Then project the quotient of s_{norm} and s_{largest} onto a score in the interval of [0; 1], which expresses the degree of similarity by subtracting it from 1:

$$\text{score}_{\text{EAI}}\left(p_i, q_j\right) = 1 - \frac{s_{\text{norm}}}{s_{\text{largest}}}$$

The best-mapping function $b_{q_j}^{p_i}$ employs a queue for edges $Q = \{e_{p_i}^1, \ldots, e_{p_i}^n\} \subseteq E_{p_i}$ that are not mapped yet, a mapping relation $M \subseteq (E_{p_i} \times E_{q_j})$ holding established mappings, a record function $R{:}E_{q_j} \rightarrow (E_{p_i}, \mathbb{R})$ storing the best angle difference found for a destination edge found so far along with its source edge, and an angle difference function $\mathrm{ad}_{q_j}^{p_i}{:}(E_{p_i},E_{q_j}) \rightarrow \mathbb{R}, (e_1, e_2) \mapsto \Delta(h_{p_i}(e_1),h_{q_j}(e_2))$.

Initially, $Q = E_{p_i}$, $M = \varnothing$, and $R(e) = (\varnothing,\infty)\forall e \in E_{q_j}$. The algorithm then repeats the following steps until $Q = \varnothing$:

1. Take one edge $e_{p_i}^k$ from queue Q so that $Q := Q \backslash \{e_{p_i}^k\}$.
2. Create sorted list $(d_1, \ldots, d_n)$ of angle differences between $e_{p_i}^k$ and each e_{q_j} incident to q_j using $\mathrm{ad}_{q_j}^{p_i}$. Also store the assignment between d_i and e_{q_j} as function ea: $\mathbb{N} \rightarrow E_{q_j}, i \mapsto e_{q_j}$.
3. For each $(d_i,\mathrm{ea}(i))$, iteratively verify record $R(\mathrm{ea}(i)) = (e_{p_i}^{\mathrm{old}}, \partial_{q_j}^{\mathrm{old}})$. If $d_i < \partial_{q_j}^{\mathrm{old}}$, enqueue $e_{p_i}^{\mathrm{old}}$ $(Q := Q \cup \{e_{p_i}^{\mathrm{old}}\})$, update $R(\mathrm{ea}(i)) := (e_{p_i}^k,d_i)$, and leave iteration (since everything that follows would be a worse assignment, as the list is sorted). Otherwise, proceed until a new difference record is found or there are no differences left.

If $Q = \varnothing$, add all projections of $R(e_{q_j}) = (e_{p_i},x)$ to M $(M := M \cup \{e_{p_i},e_{q_j}\})$, where $R(e) \neq (\varnothing,\infty)$. Then, M holds the optimum mapping and the algorithm terminates.

13.3.3.3 Exact Angular Index + Distance

It is possible to calculate a weighted score $\mathrm{score}_{\mathrm{w}}(p_i,q_j)$ that incorporates both the Exact Angular Index and the Euclidean distance with the following formula:

$$\mathrm{score}_{\mathrm{w}}(p_i,q_j) = \omega_1 \cdot \mathrm{score}_{\mathrm{EAI}}(p_i,q_j) + \omega_2 \cdot \mathrm{score}_{\mathrm{PE}}(p_i,q_j)$$

where $\omega_1 = |1 - \omega_2| \in [0; 1] \subseteq \mathbb{R}$ describes the weight given to the topological similarity of the nodes expressed by the EAI score, and $\omega_2 = |1 - \omega_1| \in [0; 1] \subseteq \mathbb{R}$ stands for the weight given to the geometrical similarity of the nodes expressed by the pure Euclidean score.

13.4 Evaluation of Point Matching Techniques

In Section 13.3, several point matching techniques were introduced. In order to evaluate these approaches, we employ an experimental setup involving real-world road maps. At first, we create a unique matching solution serving as a ground truth by manually assigning matches. Matching results of the different point matching techniques are then compared to the ground truth assignments. In this way, accuracy and performance can be measured and discussed.

13.4.1 Experimental Setup

We investigated the point matching approaches described in Section 13.2 using samples from two regions: the village of Moosach, Germany, as seen in Figure 13.1, serves as an example for relatively simple matching problems (area, 590,000 m^2; 54 nodes in reference map, 100 nodes in matching map; boundaries [48.036587, 11.870445 | 48.029227, 11.880119]), and a part of the inner city of Munich, Germany, is used as an example for difficult matching cases (area, 81,800 m^2; 26 nodes in reference map, 39 nodes in matching map; boundaries [48.151872, 11.5543 | 48.149853, 11.559203]). For the Moosach region, our sources were OpenStreetMap and a commercial map vendor, and a search radius of 40 m was set. For the sample of the inner city of Munich, we employed the ATKIS Basis-DLM map as well as OpenStreetMap data, using a search radius of 20 m due to the higher density of nodes. For each region, we manually created a ground truth matching reflecting the best association of nodes by visual inspection (Moosach, 37 matching pairs; Munich, 17 matching pairs). Each point matching algorithm was applied to each pairing of maps; then we compared the results to the ground truth matching in order to evaluate the number of *true positives* (matching pairs found in the ground truth), *false positives* (FP) (matching pairs not found in the ground truth), and *false negatives* (FN) (matching pairs present in the ground truth, but missing in the matching result generated by the algorithm). We also investigated the correlation between the score of a point matching pair and the probability of it being a true positive, in order to derive a threshold for acceptable matches. All discussed results refer to unique solutions according to Definition 13.5.

13.4.2 Results

13.4.2.1 Summary Table

The results of the evaluation are summarized in Table 13.1.

TABLE 13.1

Evaluation Results for Different Point Matching Approaches Applied to the Two Sample Regions

	Moosach (37 Ground Truth Pairs)			Munich City (17 Ground Truth Pairs)		
Algorithm	**TP**	**FP**	**FN**	**TP**	**FP**	**FN**
Pure Euclidean	26 (70%)	15	11	11 (65%)	5	6
Node valence	34 (92%)	7	3	14 (82%)	1	3
Spider Index	33 (89%)	8	4	10 (59%)	6	7
Exact Angular Index	33 (89%)	8	4	12 (71%)	3	5
EAI + distance	32 (82%)	9	5	15 (88%)	1	2

Note: TP, true positives; FP, false positives; FN, false negatives.

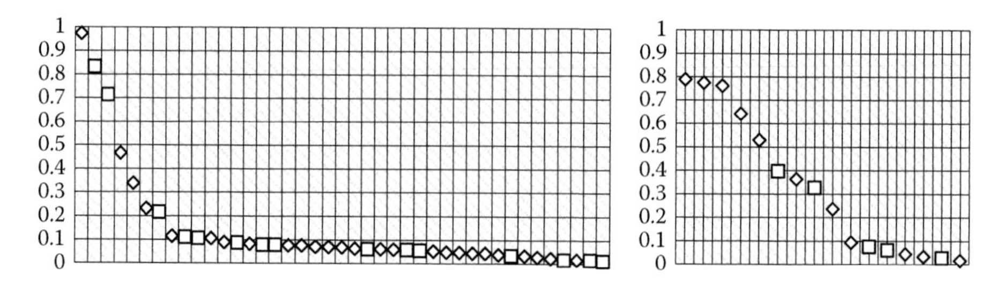

FIGURE 13.2
Pure Euclidean scores for point matches of a simple (left, Moosach) and a complex (right, Munich City) region. Diamonds, true positives; squares, false positives.

13.4.2.2 Pure Euclidean

Figure 13.2 shows decreasing matching scores of pure Euclidean matching. True positives are shown as diamonds and false positives are marked as squares. Of the 37 matching pairs defined by the ground truth matching for the Moosach sample, 26 (70%) were correctly identified. There were 15 false positives and 11 false negatives. The complex sample yielded slightly worse results: 11 (65%) correct matching pairs, 5 false positives, and 6 false negatives. Since false positives seem to be evenly distributed among the scores, a safe threshold for discarding bad matches must be set at the very end of the scale.

13.4.2.3 Node Valence

The scores of node valence matching can be seen in Figure 13.3. Node valence was able to correctly identify 34 matching pairs (92%) (7 FP, 3 FN) in the simple region and 14 matching pairs (82%) (1 FP, 3 FN) in the complex region. Clearly, node valence alone is a very coarse measure; thus, a reasonable threshold for acceptable matches cannot be established.

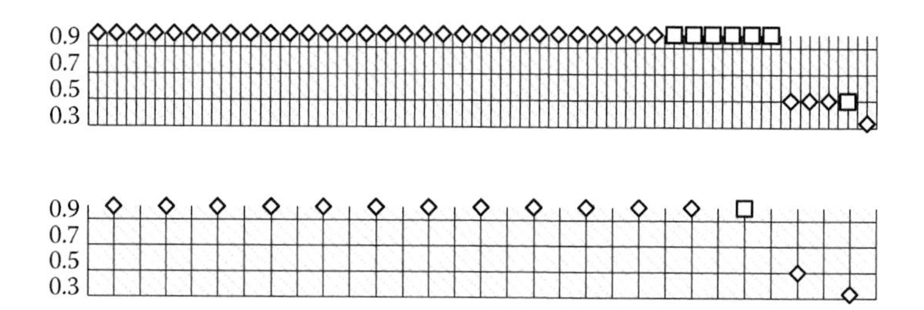

FIGURE 13.3
Node valence scores for point matches of a simple (top) and a complex (bottom) region. Diamonds, true positives; squares, false positives.

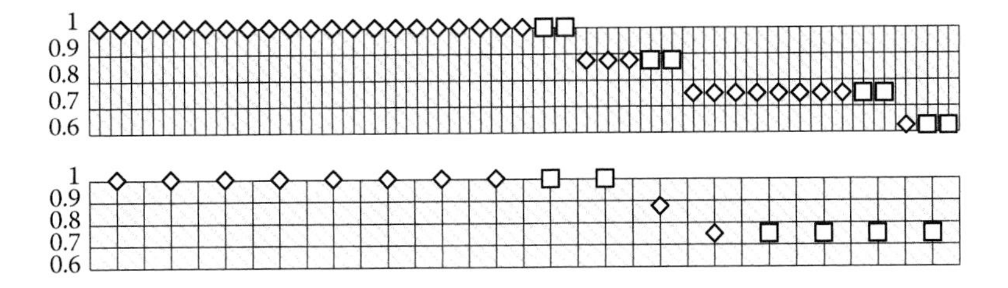

FIGURE 13.4
Spider Index scores for point matches of a simple (top) and a complex (bottom) region. Diamonds, true positives; squares, false positives.

13.4.2.4 Spider Index

The Spider Index (scores shown in Figure 13.4) identified 33 matching pairs found in the ground truth (89%) in the simple region (8 FP, 4 FN) and 10 matching pairs (59%) (6 FP, 7 FN) in the complex region. The resolution of the score is so low that for (nearly) all of the eight possible score values, true as well as false positives were found, and thus no threshold could be derived.

13.4.2.5 Exact Angular Index

Within the simple region, the Exact Angular Index (Figure 13.5) identified 33 true positives (89%) (8 FP, 4 FN), and within the complex region, 12 true positives (71%) (3 FP, 5 FN). Contrary to the Spider Index, the scores found provide a basis for establishing a threshold, as high scores are clearly, though not perfectly, correlated with true positives. In both of the samples shown, an acceptance threshold of 0.8 offers a balanced compromise that selects the most true positive matches while rejecting most false positives.

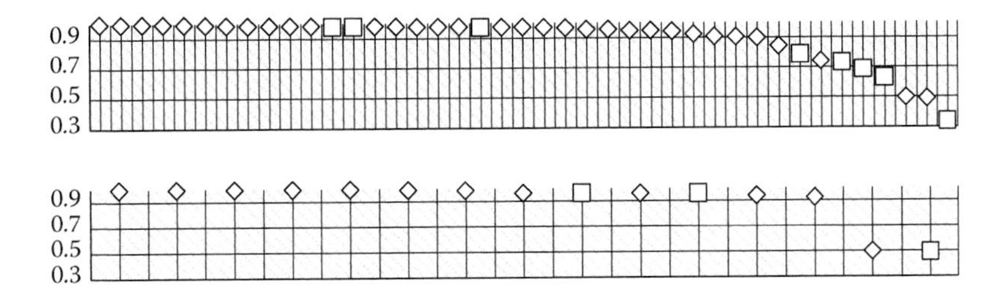

FIGURE 13.5
Exact Angular Index scores for point matches of a simple (top) and a complex (bottom) region. Diamonds, true positives; squares, false positives.

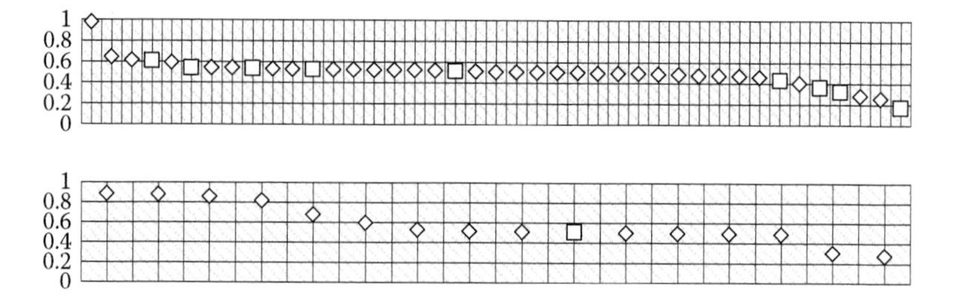

FIGURE 13.6
EAI + distance scores for point matches of a simple (top) and a complex (bottom) region. Diamonds, true positives; squares, false positives.

13.4.2.6 Exact Angular Index + Distance

The combination of the EAI score with Euclidean distance with a weight of 50% for each component yielded 32 true positives (82%) (9 FP, 5 FN) in the simple region and 15 true positives (88%) (1 FP, 2 FN) in the complex region (Figure 13.6). For the first sample, Euclidean distance deteriorates the matching accuracy of the Exact Angular Index, to an extent where a safe threshold can no longer be derived. Thus, for this sample, it can be stated that topological similarity should be preferred over geometrical distance in order to achieve good matching results. However, for the complex region, the matching result is the best of all algorithms discussed here regarding sensitivity as well as specificity.

13.5 Discussion

As can be seen in the evaluation, there is no single point matching approach that works equally well for every type of region. In general, topological and combined techniques seem to outperform matching algorithms that only rely on geometrical properties: pure Euclidean delivers mediocre results for both the urban and the rural sample, and node valence shows a surprisingly good correctness, given the fact that it does not account for geometrical distance at all. While in rural regions, any topology-aware algorithm introduced here provides reasonable results, for dense urban areas, the EAI + distance metric offers the highest correctness as well as completeness out of all evaluated approaches. Its score is also quite well, yet not perfectly, correlated with the probability of a matching pair being a correct match. In contrast, the Spider Index and node valence both are too coarse a measure to derive a threshold, and pure Euclidean does not offer a sufficient correlation at all. Further research may focus on automatically choosing an approach that works best for a certain region by investigating properties such as dispersion or density of nodes.

Point matching provides a basis for further processing steps in the process of the conflation of digital maps. In detail, the associations found between topological nodes may be employed to derive higher-level associations between composite structures such as edges, sequences of edges, or even subgraphs of the induced graph of a road network. A conflation process of high correctness and completeness enables numerous applications concerned with improving and maintaining spatial quality in digital maps, such as incremental map updating, where conflation is used to determine the differences between map versions, as well as cross-map georeferencing, which employs conflation in order to find associations between geographical objects in maps. These associations may then, for example, be used for attribute transfer or map fusion. Also, conflation provides a means for the map-agnostic representation of geographical entities, so that these can be referenced to any map of the according region offering sufficient similarity to a certain reference map.

13.6 Summary

In this chapter, we have provided an overview of different point matching techniques for road network matching. We classified and described several point matching algorithms in detail, including a novel matching algorithm called Exact Angular Index, which offers an exact metric for the topological similarity of nodes in a road network. We then presented an experimental evaluation of the point matching algorithms using real-world maps of two different regions from multiple sources. The results show that especially for complex matching cases, combinations of topological and geometrical approaches provide an advantage in both accuracy and precision, while maintaining acceptable execution times. We concluded with a discussion of ongoing challenges and applications in the field.

References

Bavarian State Office for Survey and Geoinformation, Germany. http://www.vermessung.bayern.de/service/download/testdaten/atkis.html (accessed May 6, 2014).

Christmas, W., Kittler, J., and Petrou, M. 1994. Matching of road segments using probabilistic relaxation: A hierarchical approach. In *Neural and Stochastic Methods in Image and Signal Processing III*, San Diego, CA, USA, July 24, pp. 166–174.

Gösseln, G. 2005. A matching approach for the integration, change detection and adaptation of heterogeneous vector data sets. In *XXII International Cartography Conference*, La Coruña, Spain. http://cartesianos.com/geodoc/icc2005/pdf/oral/TEMA24/Session%202/GUIDO%20VON%20GOESSELN.pdf

Kovalerchuk, B., et al. 2008. Automated vector-to-raster image registration. *Proceedings of SPIE*, 6966, 69660W-1–69660W-12. doi: 10.1117/12.778431.

Longley, P. A., Goodchild, M. F., Maguire, D. J., and Rhind, D. W. 2005. *Geographic Information Systems and Science*. John Wiley & Sons, New York, USA.

O'Donohue, D. G. 2010. Matching of image features and vector objects to automatically correct spatial misalignment between image and vector data sets. Master thesis, University of Canterbury, New Zealand.

Rosen, B., and Saalfeld, A. 1985. Match criteria for automatic alignment. In *Auto-Carto VII: Proceedings of the Digital Representations of Spatial Knowledge*. March 11–14, 1985. Washington, D.C., USA. pp. 456–462. http://www.mapcontext.com/autocarto/proceedings/auto-carto-7/pdf/match-criteria-for-automatic-alignment.pdf

Sinnott, R. 1984. Virtues of the Haversine. *Sky and Telescope*, 68(2), 159.

Walter, V., and Fritsch, D. 1999. Matching spatial data sets: A statistical approach. *International Journal of Geographical Information Science*, 13, 445–473.

Xiong, D. 2000. A three-stage computational approach to network matching. *Transportation Research Part C*, 8, 71–89.

Yuan, S., and Tao, C. 1999. Development of conflation components. Presented at International Symposium of Geoinformatics and Socioinformatics, Ann Arbor, MI, USA.

Section IV

Uncertainties in Spatial Data Mining

14

Toward a Collaborative Knowledge Discovery System for Enriching Semantic Information about Risks of Geospatial Data Misuse

J. Grira, Y. Bédard, and S. Roche

Department of Geomatic Science and Center for Research in Geomatics, Laval University, Québec, Québec, Canada

CONTENTS

ABSTRACT The aim of this research is to design a knowledge discovery system that facilitates the identification of new risks of geospatial data misuse. This research is motivated by the irregularity of risk analysis efforts and the poor semantics of the collected information about risks. As a consequence, many foreseeable risky aspects inherent to the data remain overlooked, leading to ill-defined specifications and faulty decisions. In this chapter, we present a contributed knowledge discovery system that aims to enrich the semantic information about risks of geospatial data misuse in order to identify foreseeable risks and improve its underlying quality. The proposed system relies on a systematic and more active involvement of users in risk analysis. The approach consists of (1) providing an overview of the related work, (2) presenting an ontology-based knowledge discovery system, (3) presenting the components of the proposed system, and (4) concluding with a discussion.

14.1 Introduction

In the context of geospatial database design, data quality is an issue that deserves attention in order to avoid faulty reporting and decisions while using the underlying data. Some risk-based approaches to quality assessment aim to prevent numerous types of risks, including the risks of geospatial data misuse. Those risks that may arise while using geospatial data are to be identified by experts while designing the geospatial database (Grira et al. 2010). Risk identification is a prerequisite to risk analysis and implies the use of all available information to identify intended use and reasonably foreseeable misuse (ISO 31000:2009). Geo-IT experts help, due to their domain-specific knowledge, identify such information. However, in the rare cases where this information is collected, almost no attention is paid to its representation and storage in an appropriate format that can support subsequent risk knowledge extraction. This situation leads to overlooking valuable knowledge that might be relevant for the risk analysis process.

Different stakeholders have to be involved in the risk analysis process. The literature outlines the differences in the stakeholders' backgrounds in terms of expertise and skills, a factor that may undermine the efficiency of the risk analysis process. Hence, there is a need for a common understanding of the risky issues considered. In fact, geospatial data projects need to be able to rely on a structured expert's knowledge to better analyze risks of geospatial data misuse.

The aim of this chapter is to present an ontology-based knowledge discovery system that facilitates, using a collaborative approach through a Web 2.0

platform, the identification of new risks of geospatial data misuse and the representation of the knowledge about these risks, and allows semantic reasoning on the resulting knowledge. As formal knowledge representation models, ontologies can render invaluable help in this regard. In Section 14.2, we first present an overview of the related work looking at the risks of geospatial data misuse and their management. Next, we present the concept of ontology-based knowledge representation and discovery in relation to our context of risk analysis. Then, we describe the methodology adopted to improve the knowledge on the identified risks. Subsequently, we expose in Section 14.5 the architecture of the knowledge discovery system integrated to a collaborative platform used for the collection of experts' and end users' contributions. Section 14.6 describes how the output of a knowledge discovery system can be used for geospatial database design (i.e., CASE tool integration) and decision making (i.e., web-based dashboard). Finally, we conclude in by discussing the contribution and presenting possible future work.

14.2 Risks of Geospatial Data Misuse

Nowadays, many users perceive geospatial data as being de facto reliable for their usages (Gervais et al. 2009). They usually assume that data are safe and not risky regardless of the intended usage context (Jones et al. 2013). However, perceptions about the fitness for use of the data may diverge from untrained users, geo-IT experts, and application domain experts (Grira et al. 2010; Devillers et al. 2010). Accordingly, the assumption of safe data usage has led to a number of accidents and other adverse consequences that remind us of the need to protect users against the risks of data misuse (Larrivée et al. 2011). In fact, existing geospatial technologies are known to lack effective approaches to warn end users, usually having limited expertise in the geomatic domain, of possible risks that could emerge from reusing geospatial data in a context different from the original intended use. The number of misuses involving geospatial data suggests that different sources and factors are at the origin of these misuse cases. The level of expertise of end users of the underlying geospatial data and the activities undertaken to deal with risks of geospatial misuse are both aspects that need to be analyzed in relation to the sources and factors at the origin of these misuse cases. Accordingly, we present in Section 14.2.1 the factors and sources of inappropriate use of geospatial data. Then, the users' view (i.e., perception of quality and risk) is presented in relation to inappropriate usage of geospatial data. Finally, we depict in Section 14.2.3 the different usages of geospatial data according to the expertise of the users.

14.2.1 Factors Related to Geospatial Data Misuse

The identification of geospatial data misuse cases is not a deterministic task because of the unpredictability inherent to the concept of risk. Consequently, one cannot exhaustively define a specific risk prior to its occurrence without being based on a probabilistic assessment. Nevertheless, it remains possible to describe a risk according to (1) its potential sources, (2) the factors that might trigger its occurrence, and (3) the criteria against which the significance of this risk may be evaluated. In order to describe those sources, factors, and criteria, we can refer to the domain of business intelligence (BI), more specifically to the typical data flow within its traditional global architecture (Dayal et al. 2009, 2012): in fact, a number of control points can be fixed at different steps of that architecture (Figure 14.1). We will use these points as references to describe the elements that can increase the risks of data misuse.

First, one of the main factors that can increase the risks of geospatial data misuse is the usability of the new generation of spatial BI applications. The user-friendly graphical interfaces and the ease of use of those systems allow end users to get the expected results in a few clicks instead of combining numerous technical queries. Consequently, the nature of the newly available systems motivates more users, mostly with no specific knowledge about spatial references or database structures, to use them without an appropriate understanding of the challenges and limitations of the manipulated data (Levesque et al. 2007).

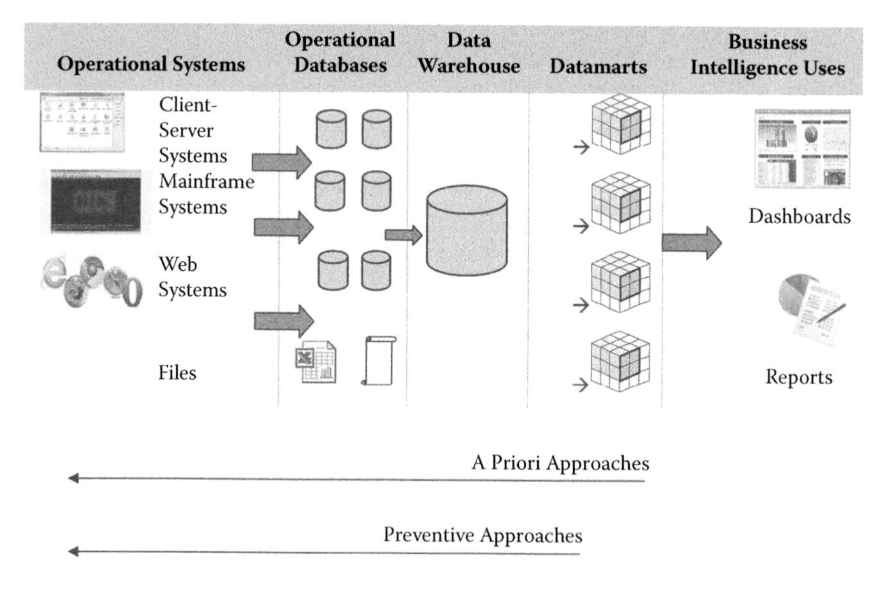

FIGURE 14.1
Traditional business intelligence (BI) architecture: data flow for decision-making process.

Furthermore, in reference to Figure 14.1, the documentation of the data (i.e., its sources and transformations) is not always properly performed, particularly at the end of the BI process (cf. datamarts, reporting); hence, the validity of the data contained in data warehouses is not systematically ensured. The ETL (extract transform load) operations applied to that data, for example, when loaded into datamarts and exported to reports for decision making, may change many aspects of the data (e.g., values, destination, format). The reliability of such data becomes more difficult to evaluate and could be undermined, potentially leading to inappropriate usages.

Finally, data integration may be another source of risk of data misinterpretation (Sboui et al. 2008). Referring to the data flow situated between the data warehouse and the datamarts and also between the datamarts and the different BI applications (Figure 14.1), many types of integration might lead to risky usages of the data. These integrations are related to the metadata, the data themselves (i.e., geospatial datamarts), and data structures (e.g., dimensions, measures, schema). Risk assessment should take into consideration such additional levels of complexity (Sboui et al. 2008).

14.2.2 Perceptions of Geospatial Data Quality and Risks

Risks may be related to a perception of the quality but can also originate from a deliberate choice. In fact, and according to Agumya and Hunter (1999), many users decide deliberately to ignore the quality information about geospatial data. This comes in response to the lack of information and the poor documentation about quality provided by data producers. In general, there is a general belief among researchers that most users disregard quality information either because they must use it or because they choose to ignore its quality (Oort and Bregt 2005).

Besides, users' perception of the usability of a data set is related to their knowledge and expertise, but also to their perception of the quality of the data to be used. In fact, information about data quality is communicated to both end users and experts with little consideration as to whether it is easily understandable and whether all spatial data consumers will have the same understanding of the provided information (Frank 1998; Boin and Hunter 2006; Goodchild 2007). With spatial data being more accessible to users with little formal training in data quality issues, the GI science community is facing a new situation that raises questions about the communicated quality (Boin and Hunter 2007) and its different users' perceptions. These perceptions correspond to the concept of fitness for use as first defined by Juran et al. (1974) in terms of "subjective concept" and described as being something that is perceived by the user. More recently, Devillers and Jeansoulin (2006a) used the same concept of fitness for use and considered that it is somewhat unique to every use case and that no single message can be communicated to all users. This concept, often called external quality, is commonly accepted in the GI quality community and corresponds to the International

Organization for Standardization (ISO) definition of *quality* (Aalders 2002; Dassonville et al. 2002; Devillers and Jeansoulin 2006b). Accordingly, reasoning about one single quality within a collaborative context is obviously misleading; the increasing number of users leads to a wide range of requirements, to different assessment processes, and consequently, to a variety of quality perceptions (Brabyn 1996; Grira et al. 2010).

In such a collaborative context, the consumers' intended use of the data and quality information vary, and so do their contributions. It would not be reasonable to think of training thousands of volunteer users, as it is unrealistic to think that geographic information providers will support all spatial data consumers. Obviously, a communication gap exists in this regard between the data provider and data consumers, as they have different perceptions of the same represented reality (Boin 2008). Completely bridging this gap is not possible because of the remaining residual uncertainty (Bédard 1987; Agumya and Hunter 2002). However, it seems possible to bring the consumer's perception of quality closer to that of the producer and to reduce the perception gap properly by using facilitating technologies that could enable such a reconciliation throughout interactive tools and Web 2.0 interfaces (Seeger 2008; Haklay et al. 2008; Scharl and Tochtermann 2007) where users may express their opinions about the provided quality. Users are still waiting for such technical facilities. Hunter (2001) also noted the absence of tools and techniques that may help to assess data quality. At the same time, data producers may take into consideration the different perceived qualities and the various possible uses of their data sets.

14.2.3 Expertise of Geospatial Data End Users

The lack of expertise of many users consists of an additional factor of risk of data misuse: users have the perception that data are safe for their intended usages, which may "deter them from adopting an informed behavior towards data" (Sboui et al. 2008). The expertise needed to perform an appropriate risk assessment may be classified according to the risk classification:

- Risks of the application domain
- Risks of inappropriate use

Experts in a specific application domain are able to assess risk related to this domain. They may not have the appropriate knowledge to foresee other types of risks related to other application domains. For example, an expert in forestry may be entitled to assess the risk of erosion on the sides of a river and the potential impacts on the vegetation. However, it might be useful to also have an expert in urban planning if some installations are planned to be built on one of the sides of the river. Both experts could cooperate to assess the risk of erosion on both vegetation (forestry) and population (urban planning). The increasing integration of many data sets originally used in various application domains makes the need of cross-domain expertise cooperation increasingly relevant.

A risk of inappropriate use of geospatial data may be considered a meta-risk. The information allowing its identification could be used in the process of decision making related to risks of a specific application domain. For example, in the case of forest fires, some details might be displayed, and others not, depending on the definition of the objects and their attributes for a given scale. Hence, especially when an aggregation is applied on the forest fire data, the operation of aggregation itself might be considered a risk factor because of the potential of an inappropriate use of the manipulated data. As a management of the identified domain application risk, reports are typically used to support decision making: the reports containing the result of the aggregation could be modified and used to prevent the occurrence of forest fires and avoid faulty decisions. However, it is the information about the type of aggregation used, that is, the identification of an inappropriate use factor, that triggered the process of domain application risk management.

A risk of inappropriate use is concerned with the study of the elements that might lead to a case of data misuse. Dealing with risks of geospatial data misuse implies identifying the factors related to the inappropriateness of use of that data. The identification of a risk belonging to this category of risks corresponds to the identification of at least an event whose occurrence might have undesirable consequences (ISO 73:2009). This event belongs to the set of possible use cases that an end user can perform on the data. In contrast, in the case of risks related to a specific application domain, the underlying events are not related to data manipulation but to the occurrence of a change in the application domain itself, leading to undesirable consequences (ISO 73:2009).

As a conclusion, risks of inappropriate use of geospatial data can require the involvement of a number of experts belonging to many application domains because of the upper-level nature of the concept of risk of geospatial data misuse and because of the increasingly frequent cross-domain projects that involve usually various types of expertise in different application domains.

14.3 Ontology-Based Knowledge Discovery

14.3.1 Ontology Definition

The notion of ontology is very useful in fields such as semantic reasoning, artificial intelligence, and knowledge management. Although there is no universal consensus on a common definition of ontology, it is generally accepted that it represents a specification of a conceptualization (Gruber 1995). Ontologies are typically defined as abstract models with a formal semantics.

Domain ontology is one kind of ontology defined as a specification of a shared conceptualization of a domain of interest (Studer et al. 1998). Domain ontologies are used to represent the knowledge for a particular type of application domain (Dittenbach et al. 2004). They represent a common formalized knowledge of a domain, as they are assumed to reflect the agreement of experts about that domain (Staab and Studer 2010; Vandecasteele and Napoli 2012).

Our interest in ontologies relies on their potential to represent and share knowledge. Ontologies help achieve a common understanding of artifacts representing human knowledge in a community, that is, the concrete representation of a model of consensus within a universe of discourse. In fact, an ontology is known as a container for capturing semantic information of a particular domain: literature outlines the usefulness of ontologies as a means to define a common vocabulary to share information in a domain (Noy and McGuinness 2001). This includes machine-interpretable definitions of basic concepts in the domain and relations among them. For the purposes of this research, the following formal ontology definition is considered:

$$O = \{C, R, I, H, A\} \tag{14.1}$$

where C is a set of concepts, R is a set of relations over concepts, I is a set of instances of concepts, H is a hierarchy of subsumption relations, and A is a set of axioms bringing constraints on Ci and Ri.

The definition presented in (14.1) converges with the widely accepted definition of ontology considering that it is "a formal explicit specification of a shared conceptualization" (Gruber 1995, p. ___) where *formal* implies that the ontology should be machine-readable and shared as well as accepted by a group or community. In the case of a domain ontology, it is usually assumed that it conveys concepts and relations relevant to a particular task or the application domain, which is the case we are interested in.

14.3.2 Ontology and Knowledge Discovery

Literature already outlined the effectiveness of using ontologies for supporting the knowledge discovery process. Evidence exists about the role of ontology in establishing correspondences and interrelations among conceptual entities in spatial and nonspatial domains (Smith and Mark 1998; Charest and Delisle 2006). Most research works addressed the issues of semantic analysis of intelligent systems (Bauer and Baldes 2005), semiautomatic reasoning (Phillips and Buchanan 2001), and other knowledge discovery mechanisms, such as concept identification, semantic relation extraction, rule acquisitions, and ontology enrichment (Marinica et al. 2010; Petasis et al. 2011).

However, none of these works have considered an ontology-based knowledge discovery process coupled to the Delphi method in order to bring end users and application domain experts within a Web 2.0 collaborative approach. They mainly focused on data representation interoperability and data interoperability. Therefore, in the context of this research, ontology-based knowledge discovery is used to extend the previously introduced collaborative risk identification and analysis process (Grira et al. 2012). It enriches a subset of the attributes of the ontology defined in (14.1). Therefore, ontology usage helps extend the collaborative knowledge repository with additional individuals, constraints, and semantic relations through the contribution of untrained users and designated application domain experts.

14.4 Methodology

The overall proposed methodology consists of five different steps: (1) ontologies, (2) risk knowledge repository, (3) semantic knowledge discovery, (4) collaborative risk analysis during geospatial database design, and (5) interpretations and usages. Each of these steps is illustrated in Figure 14.2 and explained briefly in the following sections.

14.4.1 Ontologies

This step proposes different ontologies with the goal of supporting the analysis of semantic information about the considered risks. As shown in Figure 14.2, ontologies include an (1) upper-level ontology for risks and as many as needed (2) application domain ontologies.

14.4.1.1 Risk Ontology: An Upper-Level Ontology

The first ontology outlined in Figure 14.2, the risk ontology, corresponds to an upper-level ontology. It is designed in conformance to the risk concepts defined in ISO risk standards (ISO 31000:2009). It stores concepts of risks, semantic hierarchical (e.g., is-a, part-of), and associative relations (e.g., similar-to, cause-of).

For example, as shown in Figure 14.3, the ISO concepts of likelihood and consequence of a risk are implemented as risk criteria that define the risk concept. Both of these concepts are defined in the ISO standard on risk management principles and guidelines (ISO 31000:2009).

14.4.1.2 Application Domain Ontologies

The domain ontology represents the data from the perspective of subject matter experts in that domain. This ontology relates the concepts and relations of the considered domain that the user understands. The primary

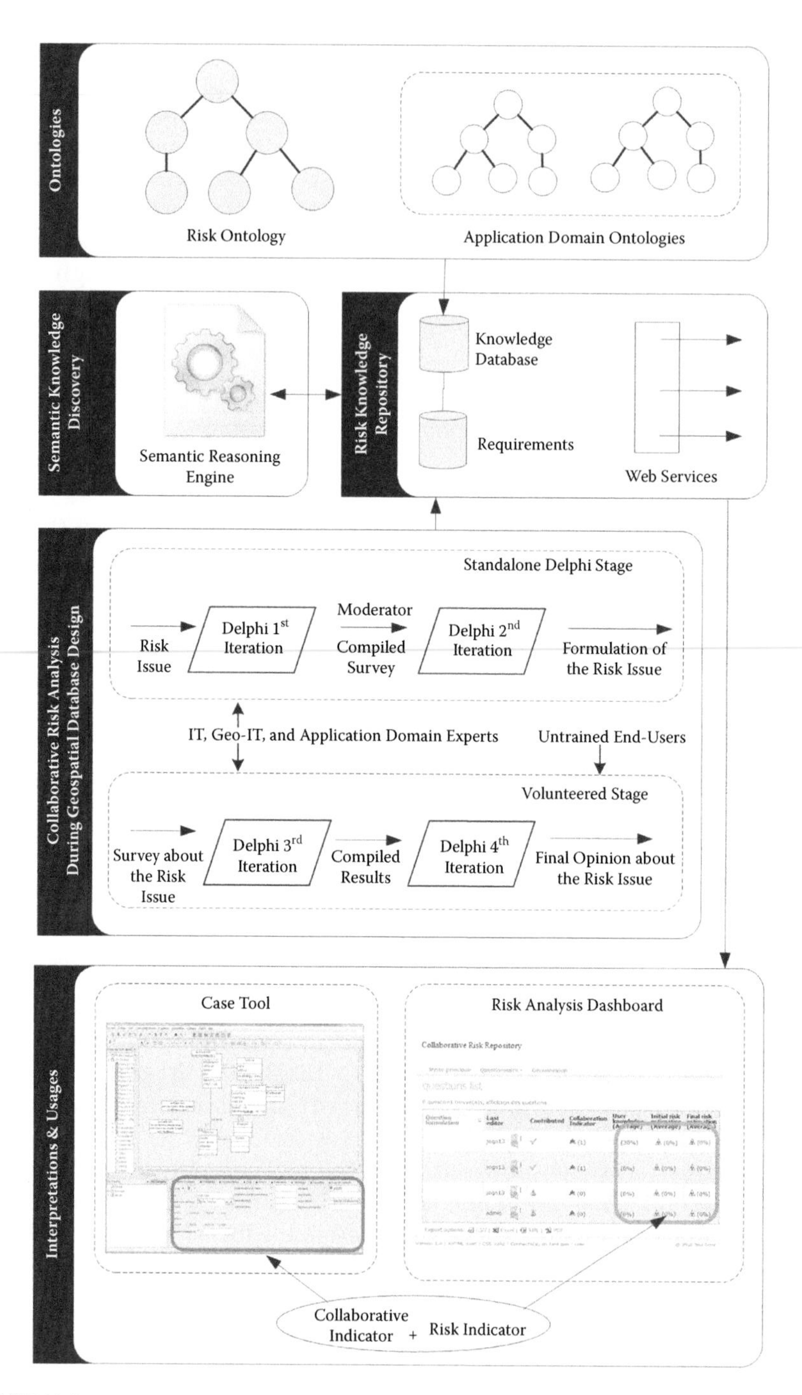

FIGURE 14.2
Enriching semantic information about geospatial data misuse: a collaborative, ontology-based approach.

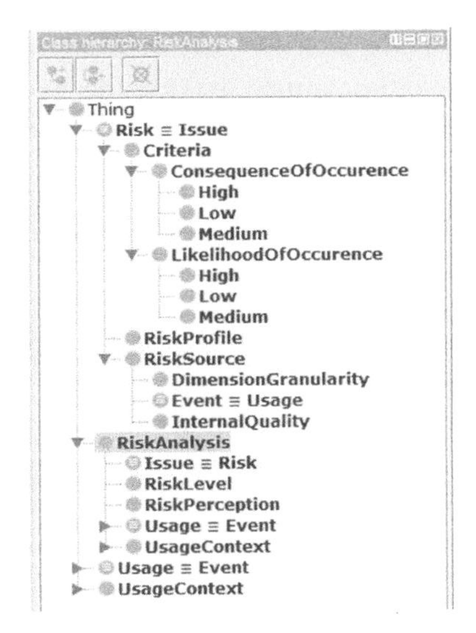

FIGURE 14.3
ISO-compliant standard risk ontology.

purpose of the domain ontology is to represent the concepts using an appropriate vocabulary for the application domain.

Application domain ontologies represent the perspective of the user community. For example, geospatial ontology should represent the knowledge of the geospatial community that uses the geospatial information. As such, it could be made up of, or derived from, other ontologies containing, for example, spatial, thematic, and temporal aspects relevant for the geospatial domain.

In the context of the present work, we conceptually represented application domain ontologies in conformance with the concept of risk profile defined in the ISO 31000 standard (ISO 31000:2009). As shown on Figure 14.3, risk profile is designed as an extendable concept that may be reused and extended to an infinite set of domain ontologies.

In relation to the geospatial domain, geo-ontologies describe the semantics of geographic data and their attributes. Literature offers a variety of contributions about the design and reuse of ontologies as a tool for system integration through the definition of a core geospatial knowledge vocabulary (Kolas et al. 2005). Sboui and Bédard (2011) proposed the use of ontologies to address the problem of semantic interoperability. Other research leveraged ontology as a semantic reference system (Egenhofer 2002; Kuhn and Raubal 2003) and as knowledge representation formalism for a particular type of application domain (Dittenbach et al. 2004)

Our work goes beyond the modelling aspects of geo-ontologies and provides a framework to utilize upper-level ontologies and domain-specific ontologies in order to efficiently address risks of geospatial data misuse that have already been collaboratively identified (Grira et al. 2012).

14.4.2 Risk Knowledge Repository

In relation to geo-ontologies defined in (14.1), the risk knowledge repository (RKR) contains the instances derived from their concepts, their relations, and their attributes; hence, according to the definition given in (14.1), the RKR may be formally expressed as follows:

$$RKR = \{I, H, A\}_{RISK} \bigcup \left(\bigcup_{k=1}^{n} \{I, H, A\}_{APPLICATION\ DOMAIN(k)} \right) \tag{14.2}$$

As illustrated in Figure 14.2, the risk knowledge repository is collaboratively enriched throughout a knowledge discovery process. This process involves end users who express their needs, intentions of use, objectives, and so forth. It also involves domain-specific experts who are responsible for identifying risks and new requirements based on their expertise and on the knowledge discovery system.

14.4.3 Semantic Knowledge Discovery

Expert knowledge and automatic, or semiautomatic, knowledge discovery have recently attracted a lot of interest. Together, expert knowledge and automatic knowledge discovery are increasingly perceived as complementary (Grira et al. 2013). In many disciplines, experts are required to provide their opinions about the system-generated knowledge (Devillers et al. 2007). The two approaches—knowledge discovery and knowledge elicitation from experts—complement rather than oppose each other.

Semantic knowledge discovery in our context consists of an expert-assisted process to elicit requirements about risks. A rule-based reasoning engine is used for basic knowledge discovery enriched with domain experts' inputs. The cooperation between expert knowledge and discovered knowledge relies on context-based interpretation of the intended usages of the data.

In our context, knowledge discovery is based on a semantic reasoning engine (Figure 14.4). However, other actors (i.e., end users and application domain experts) and components (i.e., ontologies, rules, and constraint components) help identify relevant knowledge about risks of geospatial data misuse. For example, experts are part of the knowledge discovery process, contributing with their domain-specific expertise to identify rules and patterns that may feed the risk knowledge database. Similarly, end users may contribute basic usage scenarios, which express the objectives and intentions of the use of geospatial data.

As illustrated in Figure 14.4, knowledge is input by domain experts in order to enrich the knowledge database. For example, experts may define some thresholds to be included in the rule engine: any instance that is candidate for the knowledge database should respect the predefined threshold. Otherwise, its related risks will neither be detected nor analyzed.

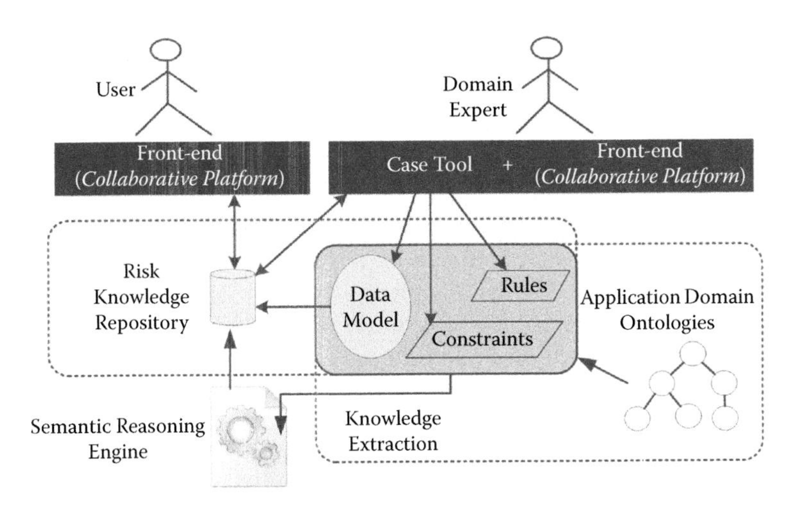

FIGURE 14.4
Semantic knowledge discovery based on expert-defined rules.

In our context, the contribution of domain experts is required in order to define domain-specific rules, identify new requirements and constraints, and assess risks related to the usage of the underlying data. In the knowledge discovery domain, literature outlines many techniques with almost no interaction with a human actor. Most of these techniques assume a clear definition of the concepts and requirements, which is often not the case. Hence, discovery knowledge techniques usually failed at incorporating valuable expert knowledge into the knowledge discovery process (Keim 2002).

Following the example of the expert-defined constraints, a threshold is considered a constraint on an instance of a concept. Constraints are types of rules defined by experts. For example, an expert-defined threshold may apply to a risk that experts decide to accept. Therefore, the threshold may be expressed as follows:

$$a \equiv po \times co \geq \delta \tag{14.3}$$

where δ is an expert-defined value, a is an axiom corresponding to an instance of A defined in (14.1), po is an instance of the concept PO, co is an instance of the concept Consequence of Occurrence (CO), Probability of Occurrence (PO) is the probability of occurrence concept defined in Figure 14.3, and CO is the consequence of occurrence concept defined in Figure 14.3.

The defined threshold is a typical example of an expert-defined rule that triggers the enrichment of the risk knowledge database. The definition of these rules is performed within a collaborative process. Section 14.4.4 describes how experts contribute to that process, especially for defining requirements for risk analysis.

14.4.4 Collaborative Risk Analysis in a Context of Geospatial Database Design

Geospatial database design is a process where at least some ad hoc activities of risk analysis have to take place. Risk analysis is important in the software design phase and constitutes a prerequisite for evaluating criticality of the system (Boehm 1991) and taking the necessary countermeasures. However, different types of expertise are required to perform the risk analysis within the design phase (Bédard 2011; Grira et al. 2012): application domain expertise (e.g., ecology, epidemiology, transportation, and security), information technologies expertise (e.g., system engineers and database designers), and geospatial information technology (geo-IT) expertise (e.g., geomatic engineers, GIS developers, and geographers). Considering the differences in perspectives, backgrounds, and objectives between the different experts, there may exist a divergence in the way they analyze risks. Accordingly, there is a need to bridge the gap between those that are experts of a domain and its requirements (i.e., application domain experts), those that are experts in the design and construction of the artifacts that together satisfy the domain requirements (i.e., geo-IT experts), and those that are experts in software and database design (i.e., IT experts) in order to get a common understanding of the risks related to the geospatial data to be used (Grira et al. 2010).

In this context, Delphi has been identified as a method that makes explicit a set of requirements and produces a collective estimation of cross-impacts of risky issues (Rowe and Wright 2011). Delphi helps end users and experts to have an exchange about design alternative solutions and arguments until a compromise, or a consensus, is reached. It provides the design team with an Agile mechanism that helps incorporate new risks on the fly into the project risk management scope (Hsu and Sandford 2007).

As illustrated in Figure 14.2, the Delphi method is not used as a standalone approach. Much research contends that using the Delphi method as only part of a wider process may well prove a means to enhance its utility (Rowe and Wright 2011). Following this research, we selected the Delphi method to involve experts in providing their judgment about the risks, but we also involved end users in a second collaborative step using a Web 2.0 collaborative platform in order to collect nonfunctional requirements (i.e., goals, objectives, and constraints) and assess their cross-impacts on risk analysis (Grira et al. 2012). Literature outlines that the use of the Delphi method with other techniques (e.g., collaborative workshops, nominal group technique, focus groups, and face-to-face meetings) usually produces more satisfactory results (Bañuls and Turoff 2011; Landeta et al. 2011) and makes them more coherent (Nowack et al. 2011).

14.4.5 Interpretation and Usages

In our context, Delphi is used in order to reach a consensus about an expert-defined threshold. In relation to the ontology definition given in (14.1), a threshold is an axiom corresponding to a constraint as expressed in (14.3). Using the Delphi collaborative process to decide the value of the threshold, the resulting set of acceptable risks may be expressed as follows:

$$r_{accepted} = \left\{ r \in RKR \middle| a \right\} \tag{14.4}$$

where $r_{accepted}$ is the set of instances of accepted risks, RKR is the risk knowledge repository defined in (14.2), r is an instance of risk in the RKR defined in (14.2), and a is the axiom defining the expert-defined constraint in (14.3).

A typical usage of the proposed approach consists of enriching an existent system used to support decision making: the decision aid system for coastal erosion risk assessment is an example (Mardkheh et al. 2012). In such a system, each risky zone is identified according to its risk probability. The latter is calculated according to many factors, such as the distance separating the road from the riverside, the soil nature, and the slope. If a threshold of 90%, for example, defines a risky road portion in the winter, this probability should not remain the same in summer. In fact, the threshold should be configurable because of the changing impact of seasons on water, soil, and the road itself. Accordingly, experts could change the probability threshold (i.e., the value of a in the RKR defined if [4]) of risky road portions as soon as they judge that the climatic conditions have significantly changed in a way to impact decision making. The decision about the effective date of the threshold change is determined through the Delphi collaborative method (Figure 14.2). Once the threshold parameter changes in the RKR, the result given by an ETL operation about the risky road portions will be different in respect to the new contributed parameter.

Risks addressed in this research are those of inappropriate usage of the data: we do not consider risks of the application domain.

14.5 Architecture

Figure 14.5 illustrates an applicative architecture for implementing the ontology-based risk knowledge discovery system. This architecture consists of three main layers: a presentation layer, a control layer, and a core layer.

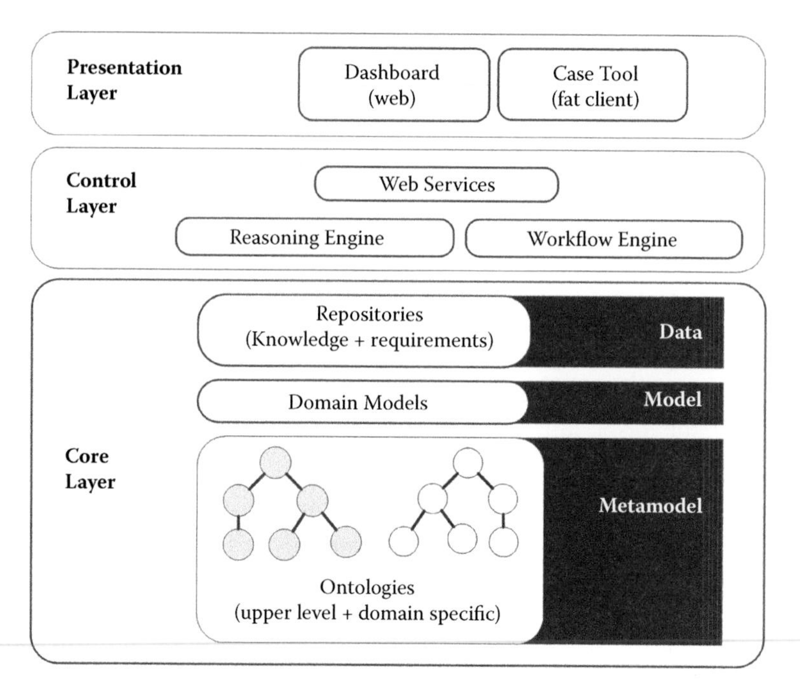

FIGURE 14.5
Architecture of the ontology-based risk knowledge discovery system.

14.5.1 Presentation Layer

The presentation layer consists on all the front ends that manage user interaction and display results to the user. In our case, the user is a geo-IT expert, an application domain expert, or a decision maker. For example, the dashboard is part of a Web 2.0 dynamic application whose role is to display indicators relevant for decision making. As illustrated in Figure 14.2, collaborative indicators and risk indicators are displayed on the dashboard: such relevant information may influence the decision-making process.

The interface of a CASE tool is also considered part of the presentation layer: once connected to the risk repository through web services, relevant information about the model to be designed is displayed. As illustrated in Figure 14.2, a "Properties" tab displays up-to-date information coming from the repositories: when an entity of the model is selected (e.g., a class, table, or attribute), the content of the tab is dynamically refreshed.

The typical users of CASE tools are IT or geo-IT experts. Knowing that some design entities are subject to risky usages may be valuable information leading designers to make important design decisions. For example, if users express their intentions of using the data for emergency operations (i.e., link between the user and the RKR in Figure 14.4), the required positional and temporal accuracy becomes critical. Besides, designers may have to change their specifications (i.e., links between the geo-IT experts and the data model, rules, and constraints in Figure 14.4) according to the newly identified intentions of use of the geospatial data.

14.5.2 Control Layer

The control layer corresponds to the layer in which business logic is implemented. In the context of our knowledge discovery system, the business logic is split into (1) a web service component, (2) a workflow engine, and (3) a reasoned engine.

Web services are exposed by the core layer for other tier systems: the CASE tool and the dashboard are typical consumers of these web services. For example, the dashboard displays risk information grabbed from the repositories: this information is displayed at the presentation layer level.

The workflow engine consists of the implementation of the Delphi collaborative process. It consists of a configurable system where surveys are built around a risky issue raised during the database design stage. The different steps of the workflows determine how many iterations users and experts should go through in order to make a design decision concerning a risky issue. New risks undetected by geo-IT experts may be identified and contributed by untrained users in this step: these risks become part of the Delphi process. Next, the result of the collaborative process becomes available and provided by the web services of the same layer.

Finally, the reasoning engine consists of the components that make up expert-defined rules and constraints and infers new knowledge based on the repository information and the experts' opinions. Many other techniques for discovering knowledge (e.g., based on similarity, rules, and patterns) may be used in order to improve the knowledge discovery processing; however, the experts' opinions remain important in our context, as illustrated in Figure 14.4.

14.5.3 Core Layer

The core layer includes (1) data, (2) models, and (3) metamodels. Metamodels represent the ontologies used in this work, that is, upper-level and domain ontologies. Domain models represent domain-specific data models. The instances of the considered ontologies and the domain models correspond to the records of the repositories. In fact, the aim of repositories is to structure and store knowledge and requirements. The structure of the knowledge repository is derived from the risk ontology and the application domain ontologies.

Some components of the proposed ontology-based risk knowledge discovery system illustrated in Figure 14.5 are presently under development. An analysis of ongoing spatial database design projects is being performed to design and implement the remaining components, particularly those of the core layer (i.e., workflow and reasoning) and of the presentation layer (i.e., connecting a CASE tool to the repositories). The Delphi-based collaborative platform and the dashboard are already in place and have been presented in previous work (Grira et al. 2012, 2013).

14.6 Conclusion

This chapter has described the design and architecture of an ontology-based knowledge discovery system that facilitates, using a Web 2.0 collaborative approach, the identification of new risks of geospatial data misuse. The risk identification is based on a collaborative process that involves application domain experts. The collaborative process is based on the Delphi method to collect experts' opinions in regards to the considered risky issues.

We then proposed an ontology-based knowledge discovery system in order to support the collaborative process and provide facilities to improve the knowledge of experts on risks of geospatial usages.

Finally, we designed an architecture that helps bringing end users and experts into the design process in a way to attenuate the risk related to potential inappropriate geospatial data usages. Nevertheless, a considerable practical challenge lies in an evolution from a one-size-fits-all risk analysis process to a collaborative approach based on the consideration of the different potential usages of the data. The adoption of our approach relies on the motivation of the CASE tool providers to offer interfaces that interact with external collaborative platforms. Our approach, based on ISO standards and widely accepted principles of software architecture, makes the integration more feasible.

References

Aalders, H. J. G. L. 2002. the Registration of Quality in a GIS. *Spatial Data Quality,* W. Shi, P. Fisher and M. F. Goodchild, Eds., London, Taylor & Francis, pp. 186–199.

Agumya, A. and Hunter, G. J. 2002. Responding to the Consequences of Uncertainty in Geographical Data. *International Journal of Geographical Information Science,* 16(5).

Agumya, A. and Hunter, G. J. 1999. A Risk-based Approach to Assessing the 'Fitness for Use' of Spatial Data. *URISA Journal,* 11(1), 1999, pp. 33–34.

Bañuls, V. A., and Turoff, M. 2011. Scenario construction via Delphi and cross-impact analysis. *Technological Forecasting and Social Change,* 78(9), 1579–1602.

Bauer, M. and Baides, S. 2005. An ontology-based interface for machine learning. In *Proceedings of the 10th International Conference on Intelligent User Interfaces*, ACM. pp. 314–316.
Bédard, Y. 2011. Data quality + risk management + legal liability = evolving professional practices. *FIG Working Week 2011*, Marrakech, Morocco, May 16–22.
Bédard, Y. 1987. Uncertainties in Land Information Systems Databases. *Proceedings of the ACSMASPRS Auto-Carto 8 Conference*, Bethesday, MD—American Congress on Surveying and Mapping, pp. 175–184.
Boehm, B. W. 1991. Software risk management: Principles and practices. *IEEE Software*, 8(1), 32–41.
Boin, A. T. 2008. Exposing Uncertainty: Communicating Spatial Data Quality via the Internet. VDM Verlag Saarbrücken, Germany.
Brabyn, L. 1996. Landscape Classification Using GIS and National Digital Databases. *Landscape Research*, 21(3), 277–300.
Charest, M. and Delisle, S. 2006. Ontology-guided intelligent data mining assistance: Combining declarative and procedural knowledge. In *Artificial Intelligence and Soft Computing*, pp. 9–14.
Dassonville, L., Vauglin, F., Jakobssosn, A., and Luzet, C. 2002. Quality management, Data Quality and Users, Metadata for Geographical Information. *Spatial Data Quality*, W. Shi, P. F. Fisher and M. F. Goodchild, Eds., Taylor & Francis, London, pp. 202–215.
Dayal, U., Wilkinson, K., Simitsis, A., Castellanos, M., and Paz, L. 2012. Optimization of analytic data flows for next generation business intelligence applications. In Topics in Performance Evaluation, Measurement and Characterization, pp. 46–66. Springer Berlin Heidelberg.
Dayal, U., Castellanos, M., Simitsis, A., and Wilkinson, K. 2009. Data integration flows for business intelligence. In Proceedings of the 12th International Conference on Extending Database Technology: Advances in Database Technology. pp. 1–11.
Devillers, R., Bédard, Y., Gervais, M., Pinet, F., Schneider, M., Bejaoui, L., Levesque, M. A., Salehi, M., and Zargar, A. 2007. How to Improve Geospatial Data Usability: From Metadata to Quality-Aware GIS Community, Spatial Data Usability. *A AGILE Pre-Conference Workshop*, 8 May, Aalborg, Denmark.
Devillers, R., Stein, A., Bédard, Y., Chrisman, N., Fisher, P., and Shi, W. 2010. 30 years of research on spatial data quality: Achievements, failures and opportunities. *Transactions in GIS*, 14(4), 387–400.
Devillers, R. and Jeansoulin, R. 2006a. Introduction, *Fundamentals of Spatial Data Quality*, R. Devillers and R. Jeansoulin, Eds., ISTE, London, pp. 17–20.
Devillers, R. and Jeansoulin, R. 2006b. Spatial Data Quality: Concepts, *Fundamentals of Spatial Data Quality*, R. Devillers and R. Jeansoulin, Eds., ISTE, London, pp. 31–42.
Dittenbach, M., Berger, H., and Merll, D. 2004. Improving domain ontologies by mining semantics from text. In *Proceedings of the First Asian-Pacific Conference on Conceptual Modelling*, Dunedin, New Zealand, vol. 31, pp. 91–100.
Egenhofer, M. J. 2002, November. Toward the semantic geospatial web. In Proceedings of the 10th ACM International Symposium on Advances in Geographic Information Systems, pp. 1–4. ACM.
Frank, A. U. 1998. Metamodels for Data Quality Description. *Data Quality in Geographic Information: From Error to Uncertainty*, R. Jeansoulin and M. F. Goodchild, Eds., Paris, Hermès Editions, pp. 15–19.

Gervais, M., Bédard, Y., Levesque, M. A., Bernier, E., and Devillers, R. 2009. Data quality issues and geographic knowledge discovery. *Geographic Data Mining and Knowledge Discovery*, 99–115.

Goodchild, M. F. 2007a. Beyond Metadata: Towards User-Centric Description of Data Quality. Keynote paper, 5th Int. Symposium Spatial Data Quality, ITC, Netherland, 13–15 June.

Grira, J., Bédard, Y., and Roche, S. 2010. Spatial data uncertainty in the VGI world: Going from consumer to producer. *Geomatica*, 64(1), 61–71.

Grira, J., Bédard, Y., and Roche, S. 2013. Revisiting the concept of risk analysis within the context of geospatial database design: A collaborative framework. ICGIS 2013. *World Academy of Science, Engineering and Technology*, 75, 970–976.

Grira, J., Bédard, Y., and Sboui, T. 2012. A collaborative user-centered approach to fine-tune geospatial database design. In *Advances in Conceptual Modeling*, 272–283. Springer, Berlin.

Gruber, T. R. 1995. Toward principles for the design of ontologies used for knowledge sharing. *International Journal of Human Computer Studies*, 43(5), 907–928.

Hacklay, M., Singleton, A., and Parker, C. 2008. Web Mapping 2.0 The Neogeography of the Geoweb. *Geography Compass*, 2(6), 2011–2039.

Hsu, C. C., and Sandford, B. A. 2007. The Delphi technique: Making sense of consensus. *Practical Assessment, Research and Evaluation*, 12(10), 1–8.

ISO 73:2009: Risk management—Vocabulary. ISO, Geneva.

ISO 31000:2009: Risk management. ISO, Geneva.

Jones, K., Devillers, R., Bédard, Y., and Schroth, O. 2014. Visualizing perceived spatial data quality of 3D objects within virtual globes. *International Journal of Digital Earth*, 7(10), 771–788.

Juran, J. M., Gryna, F. M. J., and Bingham, R. S. 1974. *Quality Control Handbook*, New York, McGraw-Hill.

Keim, D. 2002. Information visualization and visual data mining. Visualization and Computer Graphics, IEEE Transactions on, 8(1), 1–8.

Kolas, D. Hebeler, J., and Dean, M. 2005. Geospatial semantic web: Architecture of ontologies. In *GeoSpatial Semantics*, pp. 183–194. Springer-Berlin Heidelberg.

Kuhn, W. and Raubal, M. 2003, April. Implementing semantic reference systems. In *AGILE*, pp. 63–72.

Landeta, J., Barrutia, J., and Lertxundi, A. 2011. Hybrid Delphi: A methodology to facilitate contribution from experts in professional contexts. *Technological Forecasting and Social Change*, 78(9), 1629–1641.

Larrivée, S., Bédard, Y., Gervais, M., and Roy, T. 2011. New horizons for spatial data quality research. In *7th International Symposium on Spatial Data Quality (ISSDQ 2011)*, Coimbra, Portugal, 83–89.

Levesque, M. A., Bédard, Y., Gervais, M., and Devillers, R. 2007. Towards managing the risks of data misuse for spatial datacubes. Proc. of the 5th ISSDQ, Enschede, Netherlands.

Maizatul, A. I., Mashkuri, Y., and Sameem, A. K. 2006. *Ontology Construction: An Overview*. Langkawi, Malaysia. National Convention of Educational Technology.

Mardkheh, A., Mostafavi, M., Bédard, Y., Long, B., Grenier, E. 2013. Using geospatial business intelligence paradigm to design a multidimensional conceptual model for efficient coastal erosion risk assessment. *Journal of Coastal Conservation: Planning and Management*, Vol. 17, Issue 3, September, pp. 527–543.

Marinica, C., and Guillet, F. 2010. Knowledge-based interactive postmining of association rules using ontologies. *IEEE Transactions on Knowledge and Data Engineering*, 22(6), 784–797.
Nowack, M., Endrikat, J., and Guenther, E. 2011. Review of Delphi-based scenario studies: Quality and design considerations. *Technological Forecasting and Social Change*, 78(9), 1603–1615.
Noy, N. F., and McGuinness, D. L. 2001. Ontology development 101: A guide to creating your first ontology. Stanford Knowledge Systems Laboratory Technical Report KSL-01-05 and Stanford Medical Informatics Technical Report SMI-2001-0880.
Oort, P. V. and Bregt, A. K. 2005. Do Users Ignore Spatial Data Quality? A Decision-Theoretic Perspective. *Risk Analysis*, 25(6), 1599–1610.
Petasis, G., Karkaletsis, V., Paliouras, G., Krithara, A., and Zavitsanos, E. 2011. Ontology population and enrichment: State of the art. In *Knowledge-Driven Multimedia Information Extraction and Ontology Evolution*, 134–166. Springer-Verlag, Berlin.
Rowe, G., and Wright, G. 2011. The Delphi technique: Past, present, and future prospects—Introduction to the special issue. *Technological Forecasting and Social Change*, 78, 1487–1490.
Sboui, T. and Bédard, Y. 2012. Universal Geospatial Ontology for the Semantic Intraoperability of Data: What are the Risks and How to Approach Them? In T. Podobnikar and M. Ceh (Eds.) *Universal Ontology of Geographic Space: Semantic Enrichment for Spatial Data*, pp. 1–27.
Sboui, T., Salehi, M., Bédard, Y. 2008. Catégorisation des problèmes d'intégration des modèles des cubes de données spatiales, *Atellier-Fouille de données complexes—ECG'08*, Sophia Antipolis, France, pp. 1–12.
Scharl, A. and Tochtermann, K. 2007. *The Geospatial Web: How Geobrowsers, Social Software and the Web 2.0 are Shaping the Network Society*. Springer, London.
Seeger, C. J. 2008. The Role of Facilitated Volunteered Geographic Information in the Landscape Planning and Site Design Process. *GeoJournal*, 72(3–4), 199–213.
Smith, B. and Mark, D. M. 1998. Ontology and geographic kinds.
Staab, S., and Studer, R. 2010. *Handbook on Ontologies*. Springer, Berlin.
Studer, R., Benjamins, V. R., and Fensel, D. 1998. Knowledge engineering: Principles and methods. *Data and Knowledge Engineering*, 25(1), 161–197.
Vandecasteele, A., and Napoli, A. 2012. Enhancement of ontology with spatial reasoning capabilities to support maritime anomaly detection. In *Proceedings of the 7th International Conference on System of Systems Engineering (IEEE SOSE 2012).*

15

Uncertainty Management in Seismic Vulnerability Assessment Using Granular Computing Based on Neighborhood Systems

F. Khamespanah,[1] M.R. Delavar,[2] and M. Zare[1–3]

[1]*Department of Surveying and Geomatics Engineering, College of Engineering, University of Tehran, Tehran, Iran*

[2]*Center of Excellence in Geomatics Engineering in Disaster Management, Department of Surveying and Geomatic Engineering, College of Engineering, University of Tehran, Tehran, Iran*

[3]*Seismology Research Center, International Institute of Earthquake Engineering and Seismology (IIEES), Tehran, Iran*

CONTENTS

ABSTRACT Earthquakes are disasters that occur suddenly and cause enormous damage. It is vitally important to use knowledge, methods, tools, and strategies to decrease this damage, especially in urban areas. Earthquakes as recurring natural cataclysms have always been a matter of concern in Tehran, the capital of Iran, as the town is located on a number of known and unknown faults. Earthquakes can cause severe physical, psychological, and financial damage. Consequently, some procedures should be developed to assist in modelling potential casualties and their spatial uncertainty. For instance, seismic vulnerability maps should be produced as preventive measures to mitigate the corporeal and financial losses caused by future earthquakes. As vulnerability assessment is a multicriteria decision-making problem dependent on certain parameters and expert judgments, it is undoubtedly accompanied by intrinsic uncertainties. This study attempts to use a granular computing (GrC) model based on the neighborhood systems concept to address the problem of spatial uncertainty. The GrC model concentrates on a general theory and problem-solving methodology in addition to information processing by assuming multiple levels of granularity. The basic elements in GrC are the subsets, classes, and clusters of a universe. This study uses GrC to extract classification rules based on seismic vulnerability with minimum entropy to handle uncertain earthquake data. In our previous research, we implemented a GrC model based on a partition model of the universe. The model has limitations in defining similarities between the elements of the universe and in defining granules. In the model, similarities between elements are defined based on an equivalence relation. According to this relation, two objects are considered similar when they have certain attributes that are equal in value. This study establishes a general relation for defining similarities between the elements of a universe based on the neighborhood systems concept. It defines similarities based on this general relation. Instead of partitioning the universe, granulation is accomplished by covering the entire universe. This study ultimately produces a physical seismic vulnerability map of the north of Tehran based on the GrC model.

15.1 Introduction

Earthquakes are among the most well-known risks affecting urban areas. An earthquake can be described as a "vibration of the earth occurred by transmission of seismic wave from the source of elastic strain energy" (Talebian and Jackson, 2004). Although earthquakes are unavoidable, modelling earthquake vulnerability to eliminate some of the contributing parameters in high-risk regions can help decrease the number of casualties they inflict. Located on a number of known and unknown faults, Tehran, the capital of Iran, is one of the most seismically vulnerable areas in the world. According to the strategic condition of this megacity, fundamental information, in addition to vulnerability and rescue maps, is required to mitigate the effect of an earthquake event.

One of the best methods for achieving this goal involves producing a seismic vulnerability map, which generally depends on various criteria. Five effective criteria, including earthquake intensity (Modified Mercalli Intensity [MMI]), mean slope, building age, building strength, and number of building floors, are used to assess seismic vulnerability of the north of Tehran (Aghataher et al., 2005; Alinia et al., 2010; Amiri et al., 2007; Khamespanah et al., 2013; Silavi et al., 2006a, 2006b, 2008).

This study uses a granular computing (GrC) model to extract classification rules based on seismic vulnerability with minimum entropy. GrC involves the multidisciplinary study of problem solving and information processing and provides a general, systematic, and natural way of analyzing, understanding, representing, and solving real-world problems (Lin, 1997a, 2003; Yao, 2004, 2005; Zadeh, 1997; Chen and Yao, 2006). The GrC model used in this study is based on the concept of neighborhood systems to define similarities between elements according to their general relation. In this model, each element of the universe has an associated nonempty family of subsets known as a neighborhood system, and each subset is known as a neighborhood of the element (Lin, 1997b; Yao and Chen, 1997). The objective is to produce a seismic vulnerability map of the north of Tehran based on the GrC model. The area is divided into 875 statistic units.

15.2 Neighborhood Systems Concept

Sierpenski and Krieger (1956) initially presented the theory of neighborhood systems in their study of Fechet (V) spaces. The main concept of neighborhood systems is explained as follows.

For an element x that is a member of a finite universe U, a subset of U such as $n(x) \subseteq U$ may be defined. This is called the neighborhood of x. The elements in a neighborhood of x may be indistinguishable from, be similar to, or have a functional relation with x (Yao, 1998).

The nonempty family of neighborhoods of x used to explain the general kinds of relationships between the elements of a universe of x is known as a neighborhood system (Lin, 1998; Yao, 1999). A neighborhood system of x ($NS(x)$) divides the universe into classes. Different neighborhoods of x contain elements that have varying degrees of similarity to x. A neighborhood system projects every element to a family of subsets of the universe and can be considered an operator from U to 2^{2U}.

The neighborhood system of x is calculated as follows: $NS(x) = \{(x)_R\}$. If R is a reflexive and equivalence relation and obtains a reflexive neighborhood system that covers U/R, then the neighborhood system of x is the equivalent class, which is the partition U/R (Lin, 1997a).

Another kind of neighborhood system can be defined by distance (D). In this approach, U is defined as a universe function and D as a distance function. It is supposed that $D{:}U \times U \rightarrow R+$, in which $R+$ is the set of nonnegative real numbers. For every number such as d, which is a member of $R+$, the neighborhood of x (as a member of U) is defined as $nd(x) = \{y \mid D(x, y) \leq d\}$ and $NS(x) = \{nd(x) \mid d \in R+\}$ is given for a neighborhood system of x (Lin, 1997b, 1998).

15.3 Neighborhood Systems and the Granular Computing Model

This section describes the GrC model based on the neighborhood systems concept.

15.3.1 Concepts and Granules in Neighborhood Systems

Every concept is formally understood as a piece of thought comprising two parts: extension and intension (Orlowska, 1987; Wille, 1992). Extension includes items that use the same characteristics to describe the concept. In other words, it is a set of entities that provide examples of the concept. Intension contains whole attributes that are acceptable for an entire entity. Accordingly, a concept is described in terms of both its extension and intension, and this design helps to investigate formal concepts in a set-theoretic structure. Every subset of universe *U* that is like *A* can be considered an extension of specific concepts. Members of *A* are described according to specific properties that form the intension of the concept. Consider concept *A*. For each element of the universe like *x*, *NS*(*x*) is defined as follows:

$$\mathrm{NS}_{\mathrm{A}}(x)=\begin{cases}\{A\} & x\in A\\ \{\varnothing\} & x\notin A\end{cases} \tag{15.1}$$

The preceding definition means that the neighborhood system of all elements not existing in *A* is an empty set and that the set *A* is the neighborhood system of every element present in *A*. According to such a representation, set-theoretic operations can be verifiably considered neighborhood systems operations. In addition, *A* should be treated as a granule. The elements in *A* are gathered together according to their common characteristics, which means they are all examples of a definite concept.

The concepts of different GrC models are formed based on styles of definitions. This study focuses on the neighborhood systems concept and discusses the idea behind GrC as follows.

In a simple GrC model, a finite set of attributes describes a finite set of objects known as the universe, as presented in Equation 15.2 (Pawlak, 1997; Yao, 2006; Yao and Zhong, 2002).

$$S = (U, At, L, \{Va \mid a \in At\}, \{Ia \mid a \in At\}) \tag{15.2}$$

where *U* is a finite nonempty set of objects, *At* is a finite nonempty set of attributes, *L* is a language defined using the attributes in *At*, *Va* is a nonempty set of values of $a \in At$, and $Ia{:}U \rightarrow Va$ is an information function that maps an object of U to exactly one possible value of attribute a in Va.

Using this GrC model, this study applies rudimentary formulas to describe the granules and define their relationships.

15.3.1.1 Generality

The generality of concept Φ is defined as the relative size of the constructive granule of the concept, which is shown in Equation 15.3 (Yao, 2001, 2008):

$$G(\Phi) = \frac{|m(\Phi)|}{|U|} \tag{15.3}$$

where $|m(\Phi)|$ is the size of the constructive granule of concept Φ and $|U|$ is the size of the constructive granule of the universe.

15.3.1.2 Absolute Support

For two given concepts Φ and Ψ, the absolute support (AS) or confidence of Ψ provided by Φ is defined in Equation 15.4 (Yao and Zhong, 2002; Yao, 2008).

$$AS(\Phi \rightarrow \Psi) = \frac{|m(\Phi \wedge \Psi)|}{|m(\Phi)|} = \frac{|m(\Phi \cap \Psi)|}{|m(\Phi)|} \tag{15.4}$$

where $|m(\Phi \wedge \Psi)|$ is the size of the constructive granule of concepts Φ and Ψ and $|m(\Phi)|$ is the size of the constructive granule of concept Φ.

The quantity $0 \leq AS\ (\Psi|\Phi) \leq 1$ shows the degree to which Φ implies Ψ.

15.3.1.3 Coverage

The coverage Ψ provided by Φ is defined in Equation 15.5 (Yao and Zhong, 2002; Yao, 2008):

$$CV(\Phi \rightarrow \Psi) = \frac{|m(\Phi \wedge \Psi)|}{|m(\Psi)|} = \frac{|m(\Phi \cap \Psi)|}{|m(\Psi)|} \tag{15.5}$$

where $|m(\Phi \wedge \Psi)|$ is the size of the constructive granule of concepts Φ and Ψ and $|m(\Psi)|$ is the size of the constructive granule of concept Ψ.

This quantity indicates the conditional probability of a randomly selected object satisfying Φ and Ψ.

15.3.1.4 Change of Support

The change of support (CS) of concept provided by concept Φ is defined in Equation 15.6 (Yao and Zhong, 2002; Yao, 2008):

$$CS\ (\Psi|\Phi) = AS\ (\Psi|\Phi) - G\ (\Psi) \tag{15.6}$$

In this formula, $G\ (\Psi)$ may be considered a prior probability of Ψ and $AS\ (\Psi|\Phi)$ a posterior probability of Ψ. The difference between the prior and posterior probabilities is defined as the *CS*, which ranges from –1 to 1.

A positive value indicates that Φ causes Ψ, and a negative value indicates that Φ does not cause Ψ.

15.3.1.5 Conditional Entropy

Consider a family of formulas $\Psi = \{\Psi_1, \Psi_2, \ldots, \Psi_n\}$ that induces a partition $\pi\ (\Psi) = \{m\ (\Psi_1), \ldots, m\ (\Psi_n)\}$ of the universe. For formulas Φ, Equation 15.7 defines the conditional entropy $H\ (\Psi\ |\Phi)$ that shows the uncertainty of formulas Φ based on formulas Ψ (Yao and Zhong, 2002; Yao, 2008).

$$H(\Psi|\Theta) = -\sum_{i=1}^{n} p(\Psi i|\Phi)\log\left(p(\Psi i|\Phi)\right) \tag{15.7}$$

where $p(\Psi i|\Phi) = \frac{|m(\Phi \cap \Psi i)|}{|m(\Phi)|}$.

If Φ is a certain formula $p(\Psi_i|\Phi) = 1$ and $p(\Psi_j|\Phi) = 0\ \forall\ j \in 1{:}\ n$ and $j \neq i$), then entropy reaches the minimum value of 0.

In terms of the basic concepts of the GrC model, Figure 15.1 shows the granular tree as a flowchart.

15.4 Methodology

Based on their availability, the 1996 census data are used to assess the seismic vulnerability of Tehran. In addition, considering that the material and number of building floors function as effective parameters in the seismic vulnerability assessment, the percentages of materially weak buildings under and more than four stories high in any given urban district are assumed as the two major parameters (material and number of floors) affecting the seismic vulnerability. As the building design regulations in Iran were ratified and enforced in 1966, buildings constructed before this date are considered

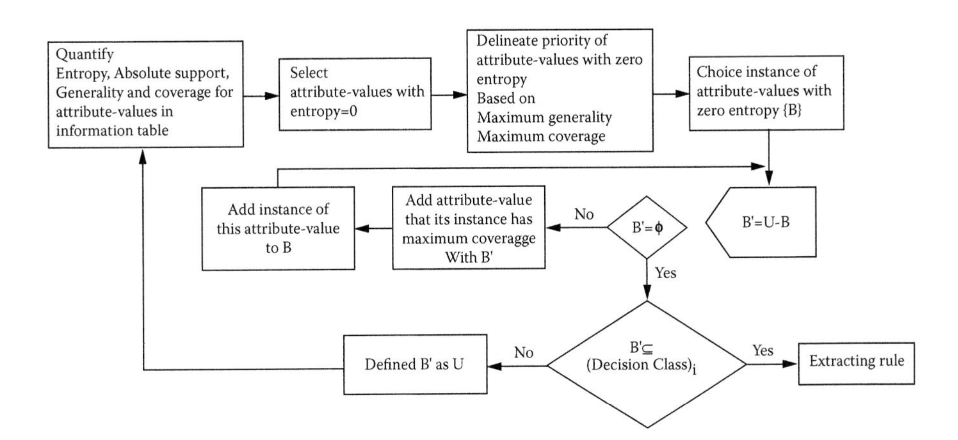

FIGURE 15.1
Granular tree algorithm for mining classification rules. (From Khamespanah, F., et al., *Granular Computing and Dempster–Shafer Integration in Seismic Vulnerability Assessment,* Lecture Note in Geoinformation and Cartography, Springer-Verlag, Berlin, 2013, pp. 147–158.)

nonstructured buildings. Moreover, the first fortification regulation (safety) against earthquakes was drawn and enforced in 1988. In this study, the two parameters of the percentages of buildings constructed before 1966 and between 1966 and 1988 are considered safety standard parameters. In addition to the building parameters, the average incline of the land and intensity of the earthquakes in the MMI are considered other seismic parameters (Aghataher et al., 2005; Alinia et al., 2010; Amiri et al., 2007; Khamespanah et al., 2013; Silavi et al., 2006a, 2006b, 2008).

Seismic vulnerability is assessed based on the activation of the north Tehran fault, and the activation of other faults is ignored. From these 875 statistic units, 30 samples are used as inputs to form a granular tree. To determine the physical seismic vulnerability of each sample urban statistical unit, experts were asked to define the physical seismic degree of vulnerability for the sample statistical units on a scale of 1–4, indicating low, medium, high, and very high vulnerability, respectively. Table 15.1 shows the 30 records selected at random. The seismic parameters are summarized as follows (Alinia and Delavar, 2011):

Slop: Slope

MMI: MMI

Les4flor: Percentage of weak buildings with four stories or less

Bef-66: Percentage of buildings built before 1966

Bet-66-88: Percentage of buildings built between 1966 and 1988

Up5flor: Percentage of weak buildings with five stories or more

TABLE 15.1

Information for 30 Randomly Selected Statistical Units in Tehran

S Num.	Slop (Degree)	MMI	Les4flor (%)	Bef66 (%)	Bet66-88 (%)	Up5flor (%)	Class
1	33.6	8.5	35	7	19	0	3
2	24.5	8.5	64	8	69	7.65	4
3	1	8.1	49	2	77	0	3
4	5.25	8.1	29	19	55	0.56	4
5	3.75	8.0	17	39	46	28.92	4
6	8.7	8.0	2	1	77	1.86	3
7	8.5	8.4	25	30	32	9.17	3
8	3.75	8.8	21	14	52	22.2	3
9	3.75	8.8	26	19	44	4.4	1
10	7.2	7.9	43	1	93	0	1
11	0.5	8.2	55	71	29	0	4
12	8.94	8.2	4	2	74	1.2	1
13	1	8.1	47	10	79	0.95	3
14	3.75	8.0	51	43	39	1.04	3
15	29.56	8.2	1	0	19	0	1
16	8.29	8.2	6	0	78	0	2
17	20.6	8.2	1	0	41	0	2
18	5.83	8.1	29	3	84	0	3
19	1	8.1	71	24	68	0.76	3
20	3.75	8.108	84	25	67	15.78	3
21	1	8.1	2	0	60	1.55	2
22	13.8	8.6	0	0	53	1.98	1
23	2.25	8.1	61	14	66	14.63	3
24	1	8	53	21	72	47.52	3
25	7.49	7.4	4	0	71	0.9	2
26	7.49	7.4	20	16	63	13.6	3
27	0	7.8	77	12	71	0	4
28	2.7	7.7	66	31	55	0.77	4
29	0	8	60	42	36	1.8	3
30	0	8.0	64	35	51	1.27	3

In the first step, the GrC model based on the neighborhood systems is used. Based on the information presented in Table 15.1, the attribute values used to construct the granular tree are considered overlapping intervals. The classes are numbered from 1 to 4, indicating low, medium, high, and very high seismic vulnerability, respectively. The intervals correspond with the attribute values as follows.

Slope		
Class	Upper Limit	Lower Limit
1	4	0
2	13	3.5
3	25	11
4	45	21

MMI		
Class	Upper Limit	Lower Limit
1	7.75	7.28
2	8.05	7.65
3	8.4	7.9
4	8.98	8.3

Percentage of Weak Buildings with Four Stories or Less		
Class	Upper Limit	Lower Limit
1	22	0
2	48	20
3	77	42
4	100	70

Percentage of Buildings Built before 1966		
Class	Upper Limit	Lower Limit
1	10	0
2	30	7
3	55	25
4	100	50

Percentage of Buildings Built between 1966 and 1988		
Class	Upper Limit	Lower Limit
1	37	0
2	60	33
3	80	55
4	100	75

Percentage of Weak Buildings with Four Stories or More		
Class	Upper Limit	Lower Limit
1	11	0
2	31	8
3	62	27
4	100	58

The concepts are defined according to their attribute value classes. The neighborhood of each element of the universe is defined based on Equation 15.1. The attribute values overlap. The granules are defined based on the neighborhood systems concept and formed based on their similarities. Any uncertainty is effectively handled in the boundaries. Similarity is a more generalized relation than the indistinguishable relation, which is used in the GrC model based on the equivalence relation. By considering the overlapping values, a neighborhood is defined based on the similarities in reflexivity and symmetry between its objects. However, these objects have no transitivity. According to this type of relation, a neighborhood of x includes x itself. For two objects x, y, if y is a member of neighborhood x ($y \in n(x)$), then x is a member of neighborhood y ($x \in y(x)$). However, if $y \in n(x)$, $y \in n(z)$, then x is not necessarily a member of $n(z)$ or $n(x)$ (Lin, 1997a).

Figure 15.2 shows the granule tree for the information presented in Table 15.1, using the GrC model based on the neighborhood systems concept.

Figure 15.3 presents the seismic vulnerability map of the north of Tehran based on the rules extracted using the granular tree in Figure 15.1.

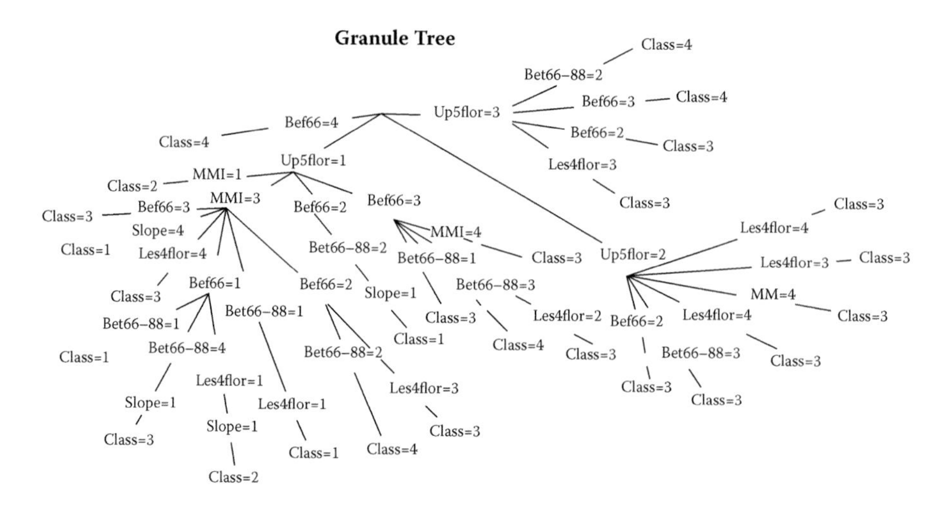

FIGURE 15.2
Granule tree of the GrC model based on the neighborhood systems concept.

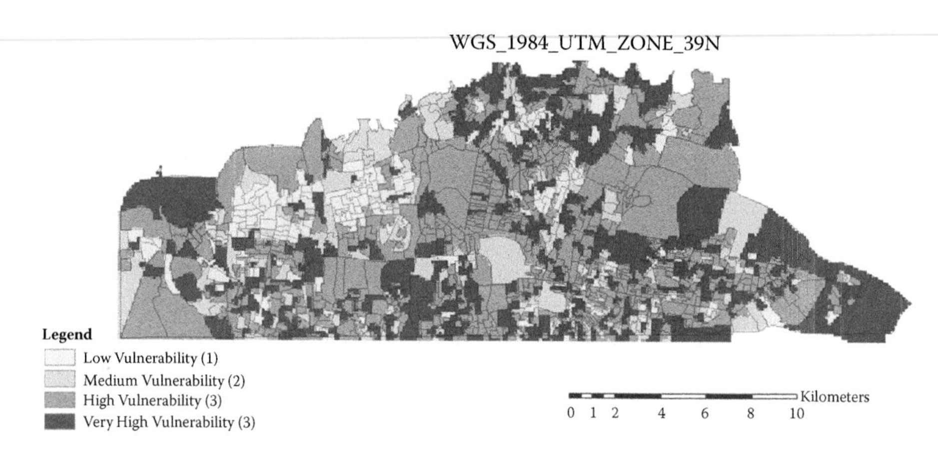

FIGURE 15.3 (See color insert.)
Seismic vulnerability map of the north of Tehran using the GrC model based on the neighborhood systems concept.

In the seismic vulnerability map provided by the granular computations, 21% of the urban units have low vulnerability, 18% of the urban units have medium vulnerability, 45% of the urban units have high vulnerability, and 16% of the urban units have very high vulnerability. According to the information presented in Figure 15.3, most of the statistical units in Tehran are classified as having high and very high vulnerability.

The accuracy of this classification of statistical units is calculated based on $k/(k + n)$ (Yao, 2004, 2005), where k denotes the correctly classified statistical units and n denotes the incorrectly classified units (Yao, 2004, 2005). Using the sample data as the test data, the final accuracy is estimated as 71%, a high level of accuracy.

15.5 Conclusion

Although earthquakes are unavoidable, modelling earthquake vulnerability to eliminate some of the contributing parameters in high-risk regions may considerably lessen the amount of casualties they inflict. Different kinds of hidden patterns can be extracted from seismic data based on the nature of the collected data, aspects of interest, and purpose of study. Many researchers have tried to determine seismic vulnerability or predict earthquakes using various models, such as partial differential equations, finite automata, and supervised learning. However, most of these approaches are unreliable due to their low levels of accuracy. This study applies a GrC model based on the neighborhood systems concept to extract rules with minimum entropy. A GrC model based on equivalence relation faces some limitations in defining the similarities between the elements of a universe and in defining granules. In this model, similarities between elements are defined based on their equivalence relation. According to this relation, two objects are similar if they share certain attribute values. This study proposes a general relation for defining the similarities between the elements of a universe based on the neighborhood systems concept. Instead of partitioning the universe, granulation is applied to cover the entire universe. This model properly addresses uncertainties in the attribute boundaries and allows more useful rules to be extracted.

References

Aghataher, R., M.R. Delavar, and N. Kamalian. 2005. Weighing of contributing factors in vulnerability of cities against earthquakes. In *Proceedings of the Map Asia Conference*, Jakarta, Indonesia, August 13, pp. 132–137.

Alinia, H., and M.R. Delavar. 2011. Tehran's seismic vulnerability classification using granular computing approach. *Applied Geomatics*, 3, 229–240.

Alinia, H., M.R. Delavar, and Y.Y. Yao. 2010. Support and confidence parameters to induct decision rules to classify Tehran's seismic vulnerability. In *Proceedings of the 6th International Symposium on Geo-information for Disaster Management (Gi4DM)*, Torino, Italy, September 15, pp. 123–130.

Amiri, A.R., M.R. Delavar, S.M. Zahrai, and M.R. Malek. 2007. Tehran seismic vulnerability assessment using Dempster-Shafer theory of evidence. In *Proceedings of Map Asia Conference 2007*, Kuala Lumpur, Malaysia, August 14–16, p. 9.

Chen, Y.H., and Y.Y. Yao. 2006. Multi view intelligent data analysis based on granular computing. In *Proceedings of 2006 IEEE International Conference on Granular Computing*, Atlanta, GA, September 23, pp. 281–286.

Khamespanah, F., M.R. Delavar, H. Samadi Alinia, and M. Zare. 2013. *Granular Computing and Dempster–Shafer Integration in Seismic Vulnerability Assessment*. Lecture Note in Geoinformation and Cartography. Springer-Verlag, Berlin, pp. 147–158.

Lin, T.Y. 1997a. Granular computing, Announcement of the BISC (Berkeley Initiative in Soft Computing) Special Interest Group on Granular Computing.

Lin, T.Y. 1997b. Neighborhood systems—Application to qualitative fuzzy and rough sets. In Advances in Machine Intelligence and Soft-Computing, ed. P.P. Wang. Department of Electrical Engineering, Duke University, Durham, NC, pp. 132–155.

Lin, T.Y. 1998. Granular computing on binary relations I: Data mining and neighborhood systems, II: Rough set representations and belief functions. In *Rough Sets in Knowledge Discovery 1*, ed. L. Polkowski and A. Skowron. Physica-Verlag, Heidelberg, pp. 107–140.

Lin, T.Y. 2003. *Granular Computing*. Lecture Notes in Computer Science 2639. Springer, Berlin, pp. 16–24.

Orlowska, E. 1987. Reasoning about vague concepts. *Bulletin of the Polish Academy of Sciences, Mathematics*, 35, 643–652.

Pawlak, Z. 1997. Rough set approach to knowledge-based decision support. *European Journal of Operational Research*, 99(1), 48–57.

Sierpenski, W., and C. Krieger. 1956. *General Topology*. University of Toronto, Toronto.

Silavi, T., M.R. Delavar, M.R. Malek, and N. Kamalian. 2006b. An integrated strategy for GIS based fuzzy improved earthquake vulnerability assessment. In *Proceedings of the Second International Symposium on Geoinformation for Disaster Management*, ISPRS, Goa, India, September 25–26, p. 5.

Silavi, T., M.R. Malek, S. Aliabady, and M.R. Delavar. 2008. Dealing uncertain spatial multicriteria system via intuitionistic fuzzy method. In *Proceedings of the Joint 4th National GIS Conference and Workshop on Decision Support System*, ISPRS, National Cartography Center, Tehran, Iran, January 12, pp. 102–108.

Silavi, T., M.R. Malek, and M.R. Delavar. 2006a. Multicriteria map overlay in geospatial information system via intuitionistic fuzzy AHP method. In *Proceedings of the 7th International FLINZ Conference on Applied Artificial Intelligence*, Geneva, Italy, August 29–31, pp. 401–408.

Talebian, M., and J. Jackson. 2004. A reappraisal of earthquake focal mechanisms and active shortening in the Zagros mountains of Iran. *Geophysical Journal International*, 156(3), 506–526.

Wille, R. 1992. Concept lattices and conceptual knowledge systems. *Computers Mathematics with Applications*, 23, 493–515.

Yao, Y. 2008. A unified framework of granular computing. In *Handbook of Granular Computing*, John Wiley and Sons, New York, pp. 401–410.

Yao, Y.Y. 1998. Relational interpretations of neighborhood operators and rough set approximation operators. *Information Sciences*, 111, 239–259.

Yao, Y.Y. 1999. Granular computing using neighborhood systems. In *Advances in Soft Computing: Engineering Design and Manufacturing*, ed. R. Roy, T., Furuhashi, and P.K. Chawdhry. Springer-Verlag, London, pp. 539–553.

Yao, Y.Y. 2001. On modeling data mining with granular computing. In *25th Annual International Proceedings of Computer Software and Applications Conference (COMPSAC 2001)*, Chicago, December 23, pp. 638–643.

Yao, Y.Y. 2004. A partition model of granular computing. *Transactions on Rough Sets I*, 3100, 232–253.

Yao, Y.Y. 2005. Perspectives of granular computing. In *Proceedings of IEEE Conference on Granular Computing*, Beijing, pp. 85–90.

Yao, Y.Y. 2006. Three perspectives of granular computing. *Journal of Nanchang Institute of Technology*, 25, 16–21.

Yao, Y.Y., and X.C. Chen. 1997. Neighborhood based information systems. In *Proceedings of 3rd Joint Conference on Information Sciences: Rough Set and Computer Science*, Research Triangle Park, NC, March 1–5, vol. 3, pp. 154–157.

Yao, Y.Y., and N. Zhong. 2002. Granular computing using information tables. *Studies in Fuzziness and Soft Computing*, 95, 102–124.

Zadeh, L.A. 1997. Towards a theory of fuzzy information granulation and its centrality in human reasoning and fuzzy logic. *Fuzzy Sets and Systems*, 19, 111–127.

16

Increasing the Accuracy of Classification Based on Ant Colony Algorithm

Ming Yu,[1,2] Chen-Yan Dai,[1] and Zhi-Lin Li[2,3]

[1]*College of Geographical Science, Fujian Normal University, Fuzhou, Fujian, People's Republic of China*

[2]*Department of Land Surveying and Geo-Informatics, Hong Kong Polytechnic University, Hung Hom, Kowloon, Hong Kong*

[3]*Faculty of Geosciences and Environmental Engineering, Southwest Jiao Tong University, Chengdu, Sichuan, People's Republic of China*

CONTENTS

ABSTRACT A new method of preliminary swarm intelligence, namely, ant colony algorithm-based classification, is investigated in this chapter. Data were collected from the regions connecting the urban and suburban areas of Fuzhou in China's Fujian Province, forming a multisource database integrating spectral information, topographical characteristics, and textural information. Classification rules were developed on the basis of differing characteristics in the samples using the ant colony algorithm. In addition, the traditional maximum likelihood method, C4.5 algorithm, and rough set method were also applied for comparison and to check the classification accuracy of our primary approach. The results show the accuracy of classification based on the ant colony algorithm to be greater than that of the other three methods. It can thus be applied to land use and land cover change in Fuzhou and other regions.

16.1 Introduction

Classification via the extraction of remote sensing (RS) data is the primary source of information for the geographic information system (GIS) in land resource applications (Pei et al., 2001; Treitz and Howarth, 2000; Yu and Ai, 2007). Mapping regional land use and cover change (LUCC) automatically and accurately from high-spatial-resolution satellite images remains a challenge. Currently, the most commonly used traditional image classification methods are based on a spectrum. However, good results cannot be obtained for many types of land use and cover based on their spectral characteristics, despite RS image classification being an important means of extracting spectral information (Lepers et al., 2005; Chen et al., 1998). Ways of boosting the productivity of such classification for use in various systems is worthy of experimentation. Spatial data mining methods are applied in many fields (D.-R. Li et al., 2002; S. Li et al., 2002; Wang, 2005; Chen and Yue, 2003). Yu and Ai (2009) compared the C4.5 algorithm and rough set method and a combination of the two, demonstrating the importance of studying LUCC information extracted using RS technology. Following in their footsteps, this chapter further discusses RS image data classification techniques based on the ant colony algorithm using different variables and compares them with other classification algorithms. The results of the study reported herein show the classification accuracy of the ant colony algorithm to be superior to that of the other algorithms tested, indicating that it should be applied in LUCC research in the future.

16.2 Study Area and Data

16.2.1 Study Area

This study was carried out in Fuzhou City in China's Fujian Province. The study area covered the region connecting the urban and suburban areas of the city, and the total grid size investigated was 512 × 512 (grid size). Geographically, Fuzhou lies between 25°15′N–26°29′N and 118°08′E–120°31′E

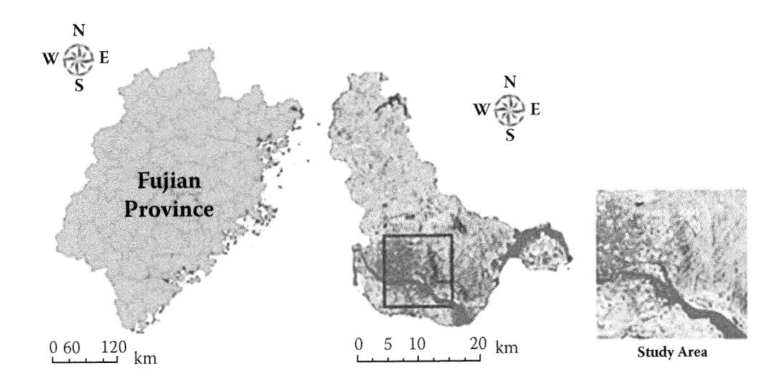

FIGURE 16.1
Study area in context.

and sits at the southeastern edge of the Eurasian continent on the southeast coast of China. This city is located on the lower reaches of the Minjiang River and on the western shore of the East China Sea. Figure 16.1 shows the study area. This region was chosen as the study area because it was the site of an earlier study, allowing more efficient comparison of the **ant colony algorithm** with the other algorithms used for classification in the same area.

16.2.2 Data Source

Some of our data came from earlier research, that is, the database in Yu and Ai (2009). The major principles we followed in classification included the principle of unity, scientific principles, applicability principles, territorial principles, systematic principles, and the monitoring principles of RS technology. On the basis of these principles and eight categories demarcating LUCC methods in conjunction with the characteristics of regional land use, the types of LUCC that connect urban and suburban regions were classified into seven categories: water bodies, wild green land, forestland, construction land, paddy fields, urban green land, and garden plots.

RS data were obtained for Fuzhou using Landsat TM/ETM+ after correction on May 4, 2000; May 10, 2003; May 8, 2007; and March 3, 2009. The space number was 119/42. As noted, the study area in which classification was carried out was a 512 × 512 grid. The bands were TM1–TM5 and TM7 and a panchromatic band with 15 m resolution, which is right at the intersecting region connecting urban and suburban areas with multiple land uses. We also used SPOT5 images of Fuzhou that had been corrected on December 14, 2003, at a multispectral band with 10 m of resolution power and a panchromatic band of 2.5 m. Finally, non-RS data were used in this study, including a regionalism map, land-use vector map, and contour vector data for Fuzhou with a scale of 1:100,000.

16.3 Method

The ant colony algorithm is a bionic technical method based on the optimization scheme introduced by Dorigo et al. (1996) and Parpinell et al. (2002). It **also constitutes a new preliminary swarm intelligence method.** Although it has been applied to RS image classification in several earlier studies (e.g., Wang et al., 2009), it is not yet in common use. Our assessment of RS image data classification techniques based on the ant colony algorithm is thus of considerable research significance. The work flowchart is depicted in Figure 16.2, and the classification steps are as follows.

Step 1: Building an LUCC classification system. In accordance with the RS image data, the recommended national standards (GB/T21010-2007) for the scheme categories are given in Table 16.1. The various LUCC categories in the study area, that is, paddy fields, garden plots, forestland, urban green land, wild green land, water bodies, and construction land, are further classified into two land types, types I and II. The visual characteristics of the land-use types in the study area are listed in Table 16.2.

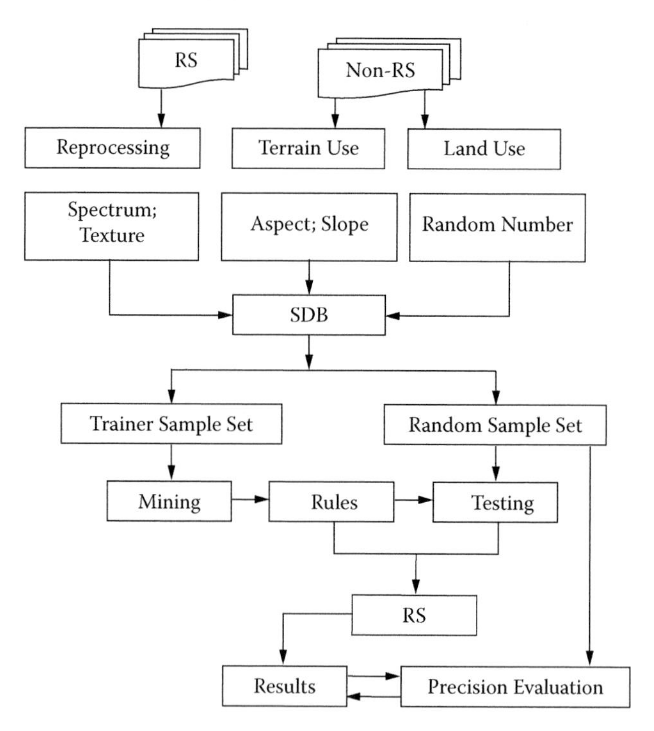

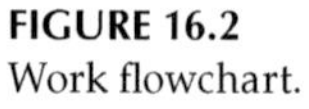
FIGURE 16.2
Work flowchart.

TABLE 16.1

Land-Use Classification Scheme for Study Area

Type I	Type II
Cultivated land	Paddy fields
Garden plots	Garden plots
Forestland	Forestland
Grassland	Urban green land
	Wild green land
Water bodies	Water bodies
Construction land	Construction land

TABLE 16.2

Visual Characteristics of Land Use in the Study Area

Surface Features	Distribution Features	Visual Image (RGB543)	Visual Image (RGB432)
Paddy fields	Rivers Housing estates	Green	Dark greenish Blue
Garden plots	Housing estates Roads	Bright green	Pink
Forestland	Hilly areas	Dark green, third dimension	Dark red, third dimension
Urban green land	Housing estates	Green, regular	Formal red
Wild green land	Hilly areas	Brownish red	Yellow-green
Water bodies	Rivers Lakes Reservoirs	Blue and dark blue, banding and irregular blocky	Blue-green and dark green, banding and irregular blocky
Construction land	Housing estates Roads Land for mining and industry	Purple	Bluish white and white

Step 2: Building a spatial database. We selected 19 correlative data entries (marked B1, B2, B3, …, B19) from early basic item databases (Yu, 2008, 2010). The classification characteristics included a digital elevation model (DEM), slope, aspect, normalized difference vegetation index (NDVI), normalized difference built-up index (NDBI), eight textural characteristics (mean, variance, homogeneity, contrast, dissimilarity, entropy, second moment, and correlation), and six gray bands (TM1–TM5 and TM7). All of these characteristics formed a decision-making table of spatial information. We then integrated multisource and multidimensional spatial data to build our spatial database.

Step 3: Classifying the ant colony algorithm based on different numbers of variables and comparing the accuracy of the results with that of other methods (see Section 16.4).

Step 4: Combining the results obtained from all four methods and comparing them across the 19 selected variables (see Section 16.4).

16.4 Comparison and Discussion

Classification accuracy in RS image classification refers to the number of pixels correctly classified. A confusion matrix is commonly used to evaluate such accuracy, the main parameters of which are producer accuracy, user accuracy, overall accuracy, omission errors, commission errors, and the kappa coefficient. Because producer, user, and overall accuracy and the kappa coefficient are all outputs in this study, we combined these parameters for result comparison.

16.4.1 Classification by Ant Colony Algorithm Based on Different Variables

In the ant colony algorithm program, decision trees are widely used for all machine learning methods (Dorigo et al., 1996; Badr and Fahlny, 2003). We applied the ant colony algorithm to obtain classification rules in the format of a decision tree (see Table 16.3) from a spatial database based on different numbers of variables (i.e., 8, 11, and 19). Figure 16.3 shows the classification results for different numbers of variables. We obtained the confusion matrix and accuracy percentages from ENVI. The accuracy percentages for 8, 11, and 19 variables were 89.12%, 90.18%, and 92.22%, respectively, and the respective kappa coefficients were 0.8651, 0.8781, and 0.9034. Figure 16.4 presents a comparison of producer accuracy based on the different numbers of variables, and Figure 16.5 provides a comparison of user accuracy.

TABLE 16.3

Comparison of Rules Based on Different Numbers of Characteristics or Variables

Variables	Rules	CF > 0.8	0.5 < CF < 0.8	CF < 0.5	Complexity
8	27	7	7	13	Simple
11	17	9	5	3	Medium
19	17	7	7	3	Complex

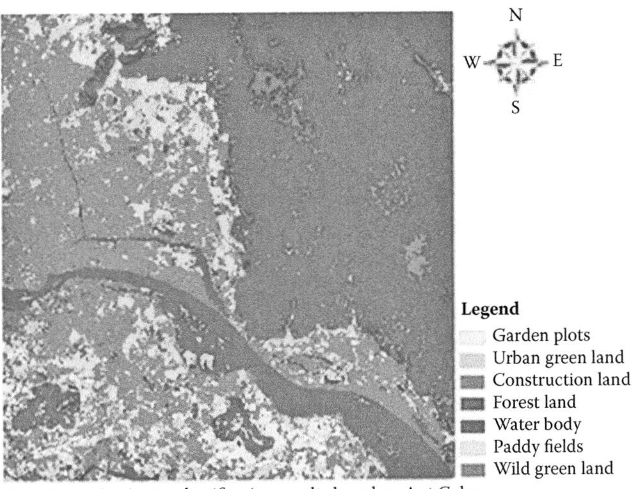

(a) Remote sensing image classification results based on Ant Colony Algorithm Optimization under the support of 8 variables

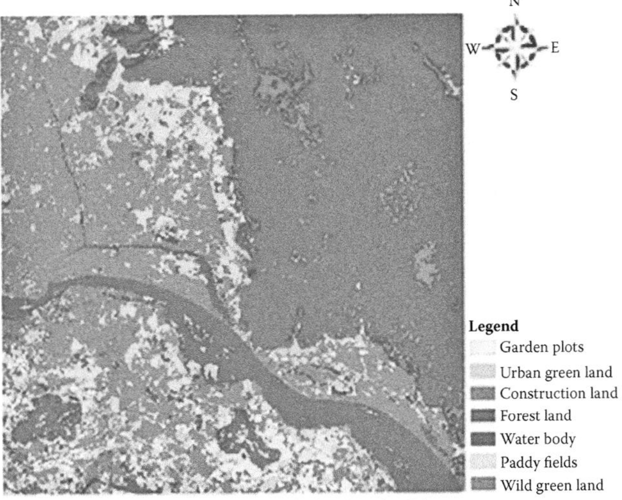

(b) Remote sensing image classification results based on Ant Colony Algorithm/Optimization under the support of 11 variables

FIGURE 16.3 (See color insert.)
Classification based on ant colony algorithm for variables a–c. *(Continued)*

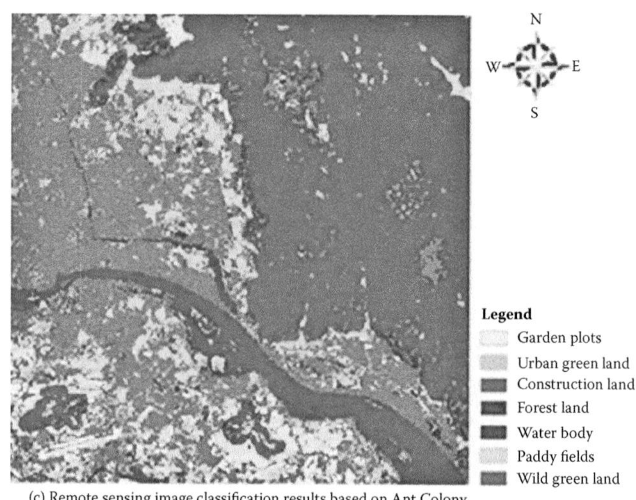

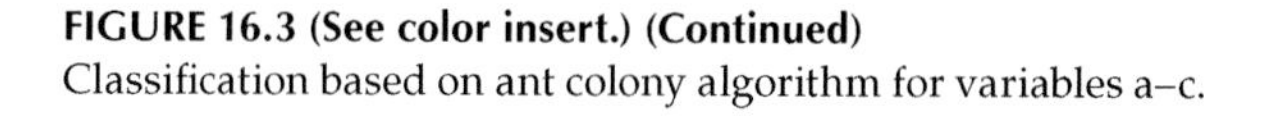
(c) Remote sensing image classification results based on Ant Colony Algorithm/Optimization under the support of 19 variables

FIGURE 16.3 (See color insert.) (Continued)
Classification based on ant colony algorithm for variables a–c.

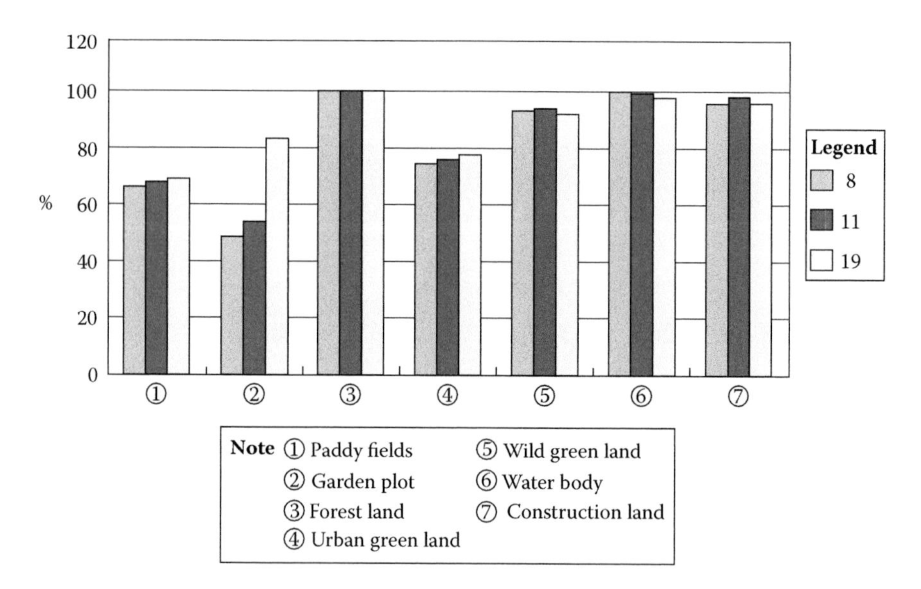

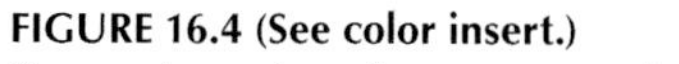
FIGURE 16.4 (See color insert.)
Comparison of producer accuracy based on 8, 11, and 19 variables.

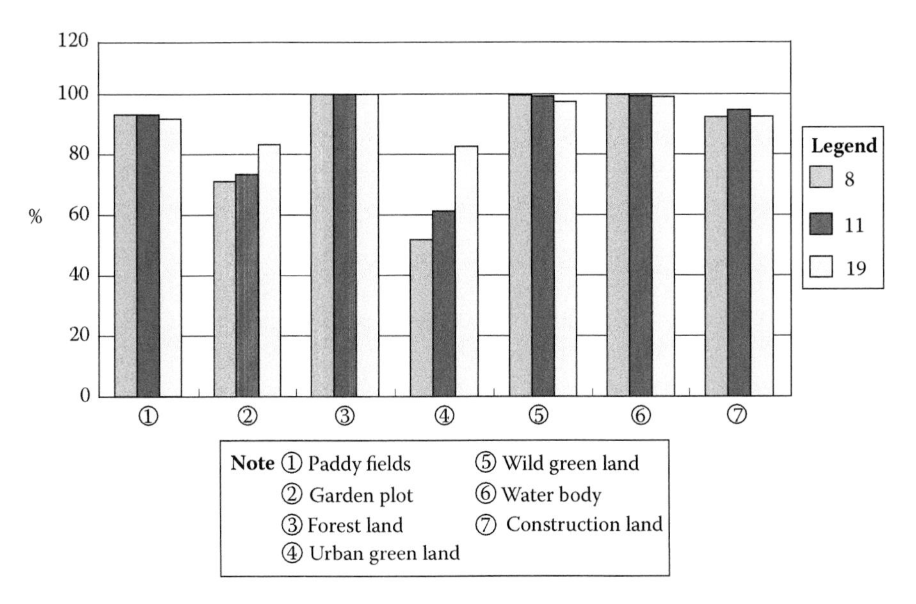

FIGURE 16.5 (See color insert.)
Comparison of user accuracy based on 8, 11, and 19 variables.

16.4.2 Select Rules Based on Different Numbers of Variables or Characteristics

A better understanding of the variables in the ant colony algorithm is an effective way of improving classification accuracy. Greater variability potentially produces more accurate results, regardless of the classification algorithm used. The two series maps in Figure 16.3 depict situations with more variables or characteristics, fewer rules, and a higher degree of complexity. Validity checks on the rules did not increase as more variables were included. Comparison results based on different numbers of characteristics or variables are presented in Table 16.3. The results are divided into three subgroups based on their complexity: the complex group includes 19 variables, the medium group 11, and the simple group 8.

Table 16.4 presents the rules used in LUCC classification of the study area (note that CF stands for classification rules) to boost confidence in the validity of the results.

Five characteristics were identified based on the experimental results:

1. As the number of variables increased, the classification results for overall accuracy and the kappa coefficient also both increased.
2. The ant colony algorithm with different numbers of variables produced generally good accuracy in RS classification. In this study, its overall accuracy was greater than 85%, with that for 19 variables

TABLE 16.4

Rules of LUCC Classification in Study Area

IF-THEN Clauses	Classification Rules
19 variables	
IF < B1 < 89.5 and B7 > 162.14 and B9 >79.81 > THEN < class = forestland 2 >	(CF = 1.000000)
IF < B9 > 79.81 and B12 < 39.5 > THEN < class = wild green land >	(CF = 0.998914)
IF < B4 < 54.5 and B12 < 39.5 > THEN < class = water body >	(CF = 0.997807)
IF < B4 < 54.5 and B8 > 148.5 and B12 <39.5 > THEN < class = forestland 2 >	(CF = 0.986413)
IF < B7 > 162.14 > THEN < class = garden plot >	(CF = 0.870421)
IF < B8 < 118.5 > THEN < class = water body >	(CF = 0.844027)
11 variables	
IF < B2 < 77.5 and B10 < 3.06 > THEN < class = forestland 1 >	(CF = 0.998873)
IF < B8 > 174.5 and 4.5 < B11 < 6.5 > THEN < class = wild green land >	(CF = 1.000000)
IF < B3 > 80.5 and B5 > 93.5 > THEN < class = construction land 1>	(CF = 0.947047)
IF < B4 < 54.5 and B9 < 48.81 > THEN < class = construction land 2 >	(CF = 0.991424)
IF < 54.5 < B4 < 74.5 and 157.5 < B8 <174.5 > THEN < class = construction land 2 >	(CF = 0.982684)
IF < B10 < 3.06 > THEN < class = paddy fields >	(CF = 0.895141)
IF < B9 > 84.33 and B10 > 3.06 > THEN < class = wild green land >	(CF = 0.836704)
8 variables	
IF < B8 > 176.5 > THEN < class = wild green land >	(CF = 0.986315)
IF < B4 < 54.5 and B5 < 63.5 > THEN < class = water body >	(CF = 0.997963)
IF < B7 < 99.89 and B8 > 176.5 >THEN < class = water body >	(CF = 0.947368)
IF < B3 > 79.5 and B6 > 81.5 > THEN < class = construction land 1 >	(CF = 0.979235)
IF < 70.5 < B2 < 89.5 > THEN < class = garden plot >	(CF = 0.863805)

standing at more than 90%, with a kappa coefficient above 0.9. The kappa coefficients were greater than 0.85 for both 11 and 19 variables. These very promising results demonstrate that the ant colony algorithm supports multiple characteristics in RS classification and is a very effective method.

3. In the three variable cases tested, the degree of accuracy was relatively high for forestland, grassland, water bodies, and land for construction, whereas paddy fields, garden plots, and urban green land generated a lower degree of accuracy because these three types of land had a mixed distribution in the RS images, and their spectra were on the borderline of the classification rules, thereby directly affecting the quality of those rules and, in turn, influencing classification accuracy.
4. Most of the cases we encountered showed a high degree of accuracy, although using 8 and 11 variables produced low degrees of user accuracy for garden plots and urban green land because the spectral

and terrain characteristics of these areas were too similar, making them difficult to differentiate.

5. The accuracy for paddy fields, garden plots, and urban green land improved considerably when 19 variables were used. The evidence collected in this study shows that use of the ant colony algorithm reduces the degree of error during the classification process, and that its results are more accurate under more complex rules for all classes.

16.4.3 Comparison of Precision Evaluation

We tested four methods of data extraction using several examples to evaluate the precision of the 19 variables (Figure 16.6a–d). The comparisons of the methods are presented in Figures 16.7 through 16.9, and their performance is discussed in this section. The following results were obtained:

1. The precision of the intelligent ant colony algorithm in RS classification is much greater than that of the traditional maximum likelihood method or rough set method, and slightly better than that

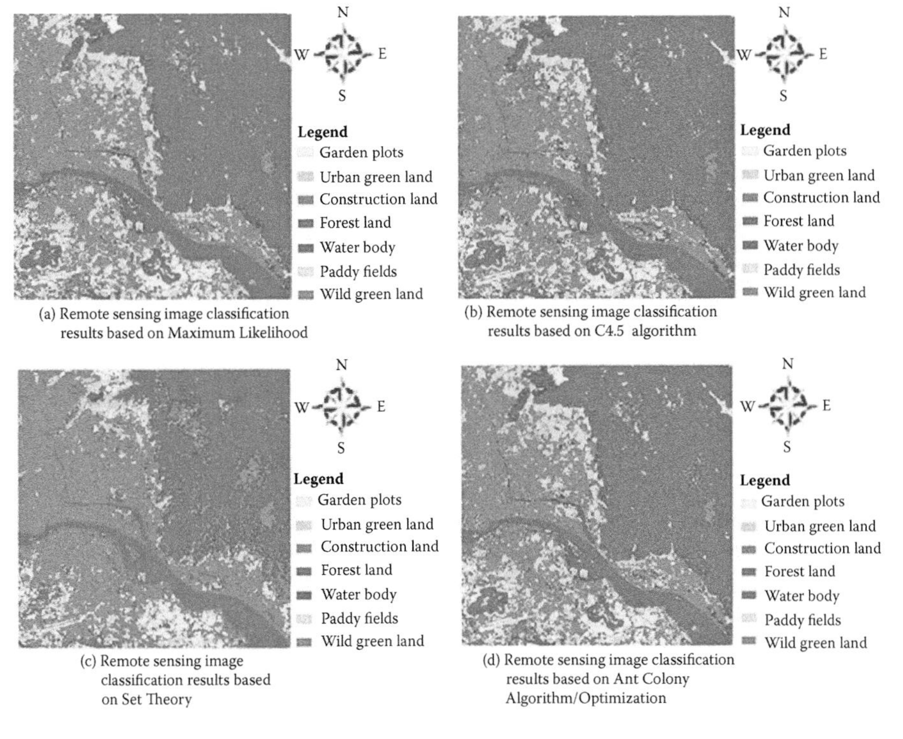

(a) Remote sensing image classification results based on Maximum Likelihood

(b) Remote sensing image classification results based on C4.5 algorithm

(c) Remote sensing image classification results based on Set Theory

(d) Remote sensing image classification results based on Ant Colony Algorithm/Optimization

FIGURE 16.6 (See color insert.)
(a–d) Different classification methods based on remote sensing images.

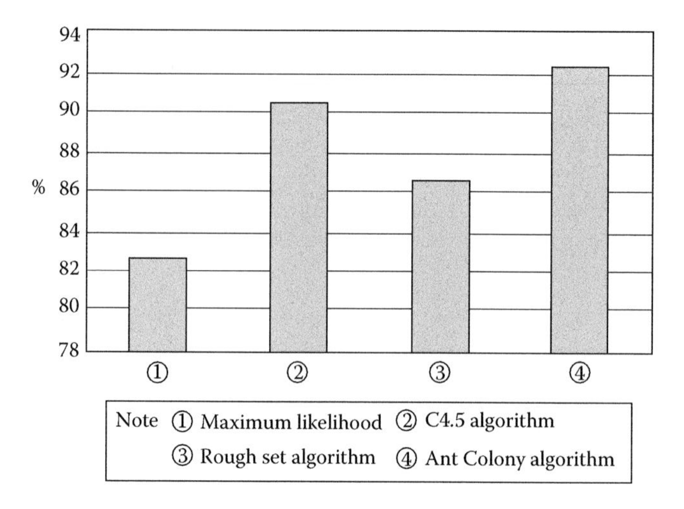

FIGURE 16.7
Comparison of overall accuracy of four classification methods.

of the C4.5 algorithm. The ant colony algorithm's overall classification accuracy was 92.22%. When compared with the maximum likelihood method, C4.5 algorithm, and rough set method, its degree of accuracy was 9.46%, 1.84%, and 5.67% higher, respectively. The kappa coefficient of the intelligent ant colony algorithm was 0.9034, with respective increments of 0.1154, 0.1154, and 0.0227 over the other three methods. See Figure 16.7 for details.

2. All four methods demonstrated better classification accuracy for regions with water bodies alone, with all achieving accuracy above 99%. The intelligent ant colony algorithm had the best classification accuracy for the other land types, followed by the C4.5 algorithm. The rough set algorithm ranked third for paddy fields, garden plots, and construction land. The maximum likelihood method generated a degree of accuracy similar to that of the ant colony algorithm for grassland, but ranked third for paddy fields, garden plots, and construction land. See Figure 16.8 for details.
3. The four methods all achieved a relatively high degree of classification precision for forestland, wild green land, water bodies, and construction land and a relatively low degree for paddy fields, garden plots, and urban green land because the spectral and terrain characteristics for the latter three land types are similar and distribution is mixed, thus influencing the selection of training data and, in turn, the classification results. The C4.5 algorithm and ant colony algorithm can both be used to reduce the influence of training data, and thus improve producer accuracy (Figure 16.8).

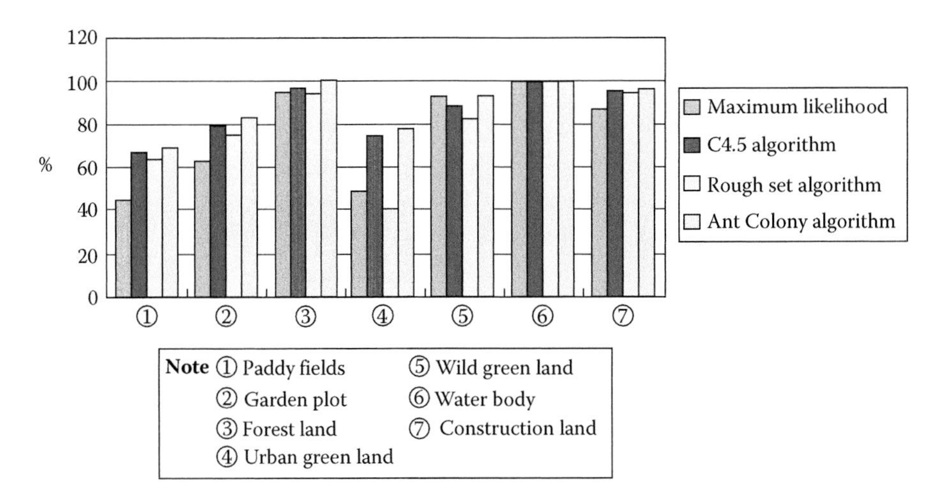

FIGURE 16.8 (See color insert.)
Comparison of producer accuracy of four classification methods.

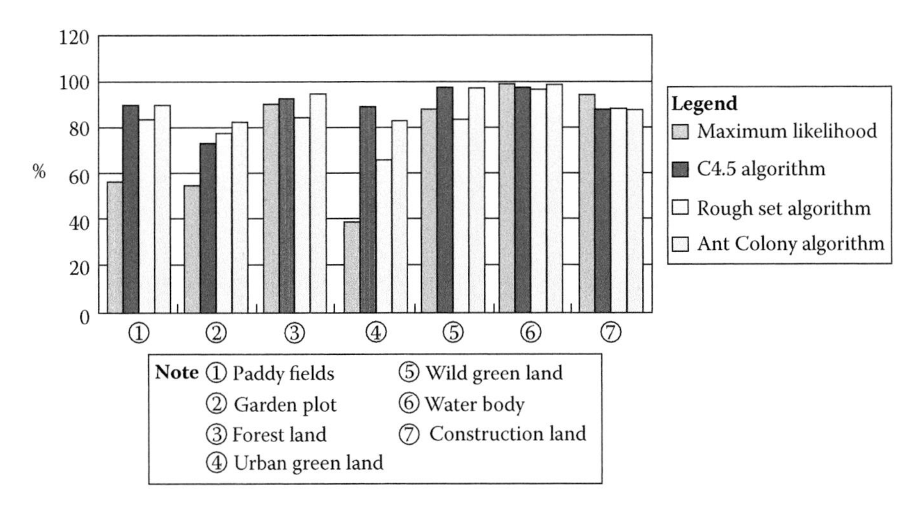

FIGURE 16.9 (See color insert.)
Comparison of user accuracy of four classification methods.

4. The results of comparison of the four classification methods suggest that land-use type classification produced a degree of precision similar to that of producer accuracy. Detailed results are presented in Figure 16.9.

In sum, the ant colony algorithm is effective for change information extraction from multitemporal image maps, as shown in Figure 16.10.

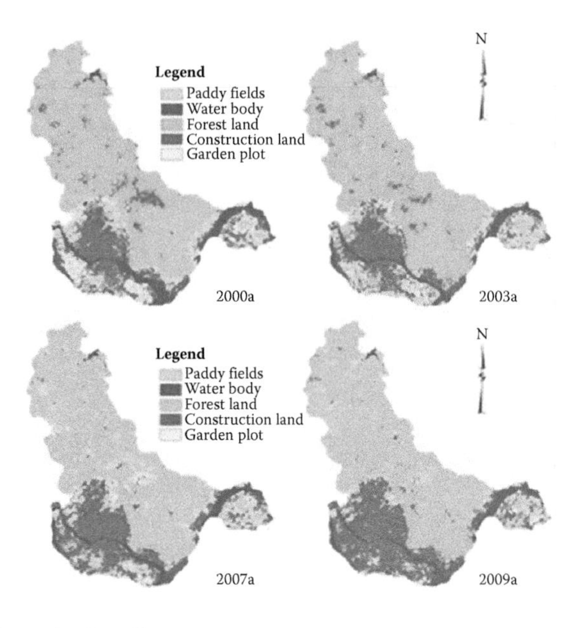

FIGURE 16.10 (See color insert.)
Classification images of Fuzhou from 2000 to 2003 and from 2007 to 2009.

16.5 Conclusion

This chapter presents a useful RS classification scheme based on investigation of the ant colony algorithm's application to the classification of remotely sensed images. The chapter also explores the effectiveness of that algorithm for multiple features of RS classification compared with three other widely used schemes. The accuracy of the ant colony algorithm has been verified, and the method has been shown to produce more precise classification results than other methods. The proposed scheme is thus recommended for use in regional LUCC. It is the fastest and most effective approach currently available for environmental monitoring using multitemporal RS images.

Acknowledgments

This work was supported by research funding from Fujian Province (2013R02 and 2014J01149).

References

Badr A, Fahlny A. A proof of convergence for ant algorithms. *International Journal of Intelligent Computing and Information*, 3(1):22–32, 2003.

Chen S-P, Tong Q-X, Guo H-D. *Mechanism of Remote Sensing Information*. Science Press, Beijing, 1998.

Chen Z-X, Yue C-Y. Research and development on spatial data mining. *Computer Engineering and Applications*, 39(1):5–7, 2003.

Colorni A, Dorigo M, Maniezzo V, et al. Distributed optimization by ant colonies [C]. In *Proceedings of the 1st European Conference on Artificial Life*, 1991, pp. 134–142.

Dorigo M, Maniezzo V, Colorni A. Ant system optimization by a colony of cooperating agents. *IEEE Transactions on Systems, Man, and Cybernetics Part B*, 26(1):29–41, 1996.

Lepers E, Lambin EF, Janetos AC, et al. A synthesis of information on rapid land-cover change for the period 1981–2000. *BioScience*, 55(2):115–124, 2005.

Li D-R, Wang S-L, Li D-Y, Wang X-Z. The theories and methods on spatial data mining and knowledge discovery. *Journal of Wuhan University* (Science Edition), 27(3):221–233, 2002.

Li S, Ding S, Hu S. Comparative study on remote sensing image classification method. *Journal of Henan University* (Natural Science Edition), 32(2):70–73, 2002.

Parpinell RS, Lopes HS, Freitas A. Data mining with an ant colony optimization algorithm. *IEEE Transactions on Evolutionary Computation*, 6(4):321–332, 2002.

Pei T, Zhou C, Han Z, Wang M, Qin C, Cai Q. Progress review on spatial data and knowledge discovery. *China Image and Graphics Journal*, 6(9):854–860, 2001.

People's Republic of China. Land-use status classification [S]. National Standard GB/T21010-2007. August 8, 2007.

Treitz P, Howarth P. Integrating spectral, spatial and terrain variables for forest ecosystem classification. *Photogrammetric Engineering and Remote Sensing*, 66(3):305–317, 2000.

Wang H-Q, Wang J-F. Research progress on spatial data mining technology. *Geography and Geographic Information Science*, 21(4):6–10, 2005.

Wang S, Yang Y, Lin Y, Cao C. Automatic classification of remotely sensed images based on artificial ant colony algorithm. *Computer Engineering and Applications*, 41(29):77–88, 2005.

Yu M. *The Generations and Applications of Synthetic Geo-Information TUPU for Ecosystem [M]*. Surveying and Mapping Press, Beijing, 2008.

Yu M. *Study on Synthetic Geo-Information TUPU for Urban Heat Environment Based on RS Images [M]*. Surveying and Mapping Press, Beijing, 2010.

Yu M, Ai T-H. Study on water body extraction and wetland sorts based on SPOT5 images. In *Proceedings of the Third International Symposium on Intelligent Earth Observation Satellites*. Science Press, Beijing, 2007, pp. 239–243.

Yu M, Ai T-H. Data mining and C4.5 algorithm and its classification application. *Proceedings of SPIE*, 7492, 74920B-1, 2009.

Yu M, Peng Y-R, Ji Q. Study on urban thermal environment based on RS and GIS techniques: Taking as example in coastal cities of southeast Fujian Province. Presented at IEEE International Conference on Electronics, Communications and Control (ICECC), 2011.

Index

S

T - #0174 - 311019 - C12 - 234/156/14 - PB - 9780367377144